Dominate Your Market!

The Attorney's Complete Guide to Online Marketing and Social Media

By Attorneys

Robert Armstrong & Sanford M. Fisch

ISBN 978-0-9841215-4-0

Contents

Introduction i
How to Use This Book iii
Other Books by the Authors v

Part 1: Gearing Up For Change
Chapter 1: That Was *Then,* This Is *Now* 3
Chapter 2: The Microchip Revolution 5
Chapter 3: Why Traditional Marketing Won't Work 6
Chapter 4: The Evolution of the Attorney 7
Chapter 5: Why You Need an Online Presence 8

Part 2: On Your Mark
Chapter 6: What Is "Online Marketing"? 11
    Inbound vs. Outbound: What's the Difference? 11
    Research and Development 16
    Promotion 16
    Analysisa nd Adjustment 16
  6.1: The Problem(s) with Online Marketing 17
Chapter 7: Missions, Visions, and Branding: What Does It All Mean? 18
  7.1: Do You Have A Vision? 19
  7.2: Identifying Your Ideal Client 20
  7.3: What Happens When Your Vision Changes? 20
    How Do You Know When You've Chosen the Right Path? 21
  7.4: Getting the Right Mindset 21

Part 3: Get Set...
Chapter 8: Learning the Basics 25
  8.1: A Crash Course in Internet Marketing Lingo 25
  8.2: A Few Thoughts on Domain Names 26
    .Com or .Net? What Extension Should You Choose? 27
    Keep It Short (And Simple) 28
    How Much Does It Cost? 28
    The Relationship Between Domains and SEO 29
    Subdomains vs. Folders vs. Separate Sites 29
    Avoiding Duplicate Content 30
  8.3: What You Need to Know About Hosting 31
    Types of Hosting Accounts 31
    Choosing the Right Hosting Account for You 33
Chapter 9: Everything You Need to Know About SEO 35

9.1: The Three Primary Components of SEO **36**
   Step# 1—Keywords **37**
   Step #2—Optimizing Your Site **43**
   Step #3—Building Backlinks **52**
   Final Thoughts on the SEO Process **58**

9.2: Local vs. National Search **58**
   Building Local Backlinks **59**
   Google Places **59**
   Yahoo! Local **65**
   Bing Local **67**
   Focus on Relationships **68**
   Getting Your Clients' Reviews **69**
   3 Principal Ways to Boost Local Search Results **70**

Chapter 10: Setting Up Shop **71**

10.1: Taking Stock **71**

10.2: Choosing Your Site Design **72**
   5 Things Every Website Should Have **72**
   Choose Your Graphics Carefully **73**

10.3: The Beauty of Blogs **73**
   How WordPress Works **76**
   Posts vs. Pages **77**
   Categories vs. Tags vs. Keywords **77**
   Configuring WordPress to Name Your Posts **78**
   Choosing Your Blog's Design **79**
   Creating Your First Post **79**
   How Often Should I Blog? **82**
   What Should I Blog About? **84**
   The Importance of Creating Quality Content **86**

10.4: Creating Marketing Materials that Really Work **86**

Part 4: Dominate!

Chapter 11: Establishing Yourself as the Expert **91**

11.1: Blogging **91**
   Blog Feeds **92**
   Blog Pings **93**
   Social Bookmarking **94**
   Digg (PR8) **95**
   Importing Your Blog **97**
   Building Followers on Digg **98**
   Mixx (PR8) **98**
   StumbleUpon (PR8) **103**
   How to Find People to Follow on StumbleUpon **105**

11.2: Article Marketing **106**
   Ezine Articles **106**
   Article Dashboard **106**

Squidoo **107**
Hub Pages **109**
Helium **112**
Associated Content / Yahoo! Contributor Network **118**
What Should I Write About? **120**
How Often Should I Publish? **121**

11.3: Landing Pages and Special Offers **121**
Landing Page Structure **121**
Landing Page Usage **125**
Separate Domain or Not? **125**

11.4: Videos **126**
Video Styles **127**
Video Dos and Don'ts **128**
Submitting Your Videos **129**
How to Get the Most from Video Marketing **131**

11.5: Podcasting **132**
How To Create Your Own Podcast **132**
Podcast Recording Software **132**
Podcast Promotion **133**

11.6: Mobile Marketing **134**

11.7: Press Releases **136**
Writing Your First Press Release **136**
Making Your Press Release Social **138**
Avoiding the Marketing Hype **139**
What's Considered Newsworthy? **139**
Distributing Your Press Release **139**

Chapter 12: Using Social Media to Build Your Community **141**

12.1: What Is Social Media? **141**

12.2: Different Types of Social Media **141**

12.3: Getting Started With Social Media **143**
Facebook **143**
Setting Up Your Personal Facebook Profile **144**
Separate Identities: Should You Maintain Multiple Facebook Accounts? ... **144**
Building Your Firm's Fan Page **145**
Facebook Groups **150**
How To Increase Friends/Fans/Followers on Facebook **151**
Twitter **152**
Setting Up Your Twitter Profile **152**
What's a Tweet? **154**
What Should I Tweet About? **156**
How To Build Your Following on Twitter **157**
Great Twitter Tools **159**
LinkedIn **160**
Getting Started With LinkedIn **161**
Building Your Network **162**
MySpace and All the Other SM Sites **163**

12.4: A Quick Word on Ethics **164**

Chapter 13: The Big *Don'ts* of Online Marketing **165**

Mousetrapping **165**
Spamming... **165**
Keyword Crazy.. **165**
Simple Is Best... **166**
Link Baiting **166**

Part 5: Bringing It All Together

Chapter 14: Analyzing Your Results **169**
Chapter 15: Keep the Momentum Going **174**
Chapter 16: Do You Need a CMO?.. **175**
Chapter 17: What You Need to Know about Outsourcing............................ **176**
Exclusive Offer from Best Legal Practices **178**

Appendices

Legal Directories.. **180**
Local Directories **182**
Search Engines.. **184**
Blog Directories.. **185**
URL Shortening Services **186**
Email Marketing Services... **187**
Social Networking Sites.. **188**
Social Bookmarking Sites... **189**
Video Sharing Sites **190**
Podcast Hosting Sites... **191**
Internet & Online Marketing Lingo **192**
About the Authors **195**
Robert Armstrong... **195**
Sanford M. Fisch **196**
Notes.. **198**
Notes.. **199**
Index **200**

Introduction

You are about to dive into the world of online marketing.

And, if you're like most attorneys, you're probably feeling a little anxious about the whole thing. You're excited about the prospect of bringing in new clients and adding profits to your bottom line, but you also know there's a lot you'll need to learn and that requires time you simply don't have.

Besides, where do you start?

There are so many components to online marketing, many of which you're not even sure you need. After all, you've spent your entire career doing just fine without having to go online at all, so why should you now have to worry about things like social media and blogging to bring in new clients? Maybe just having a website would be enough. Maybe learning a little bit of SEO would do the trick.

We understand. We know that feeling too. When we first started looking at the concept of online marketing way back when, we realized that there was an entire world out there that we didn't even know about. And why would we? We're attorneys!… we don't spend our time surfing the web… we practice law.

But training and educational resources are a big part of the service that we offer our members at the American Academy of Estate Planning Attorneys and in order for us to ensure that our service was *"cutting-edge"*, we had to immerse ourselves in the online world.

What we found is that having a website and learning a little SEO is just scratching the surface. In order to dominate online, you have to really understand what online marketing is: a way of connecting with your target market—that is, your ideal client—and then persuading those prospects that you're the right attorney for the job.

To do this, you'll have to participate in the entire process because—and we know this from personal experience—the process changes quickly. Just when you've got your website the way you want it, new technology is going to suggest you do it differently. Right when you think you've got the whole search engine thing figured out, the search engines will change the way they rate your site.

What's more, this process doesn't really revolve around your website anymore. It revolves around your clients. This is why you're hearing so much buzz about social media. The entire web has gone social.

So, the idea that you can skip some of the courtship and go right to the contract is a dangerous one. Your prospects are looking for a certain kind of relationship and, if you can't give it to them, they'll go looking somewhere else.

Now, if you were a regular Internet business, that might be the end of this conversation. But you're not… and it isn't.

In addition to this general need to join the online marketing movement, we have to modify all these strategies and techniques because—well, let's face it—running a law practice is unique.

Our clients are different. Their needs are different. And, the relationship we build with them is different. Because of that, we can't just *"turn our website on autopilot"* as many of the Internet marketing companies advise. Group coaching and web-based membership sites are out of the question in our line of work. So, many of the sales funnels and call-to-action techniques you'll see recommended on the Internet will have to be seriously tweaked to fit our purpose.

So, while we know you want to streamline the process and get to the end result, we're going to suggest (strongly) that you don't do that. And here's why:

Launching a marketing campaign is akin to throwing a pebble into a pond. It causes a ripple effect which can extend out well beyond where the pebble actually landed. In terms of marketing, that pebble represents your first few efforts in the online world. And while building a lasting and respected online presence takes time, those first few efforts do, typically, generate some sort of response. You want to be sure you create the right impression the first time.

Maybe you get a few extra inquiries regarding your firm and the services you offer. Maybe you see an initial boost of followers to your social media accounts, or maybe your website traffic increases noticeably over the next few weeks.

Whatever the case, that initial boost can be a springboard for a continuous stream of new activity if you've planned your campaign correctly.

But planning requires preparation and when you skip steps, you're essentially sabotaging your own efforts. You won't have the processes in place to keep the momentum going once it's started and, the result is that you'll lose more ground than you gained.

Those new visitors and followers are intrigued. They're waiting to see what you're going to do next. If you aren't ready to start blogging daily, for example, or if your first *"tweet"* is followed by silence for the next several weeks, you're essentially telling your new audience that you can't be bothered. It's like promising the jury you have an iron-clad case and then not even bothering to show up for trial—and we both know that's not a strategy you'd follow.

So don't do it here either. Follow the steps as they're laid out and, by the time you reach the launch phase of the process, you'll really be ready. And when you discover that the process has changed yet again or that you need to modify a strategy to make it work for you, you'll be ready for that, too.

And just between us, preparation is half the battle in dominating your market.

How to Use This Book

The ideas and strategies outlined in this book are based on generally-accepted marketing practices however, different states have different guidelines about what attorneys can and cannot do when it comes to advertising.

Because of that, we cannot assume any responsibility for penalties, reprimands or other adverse action taken against you (or your firm) in response to marketing activities initiated by you or others in your firm as a result of reading this book. To avoid such penalties, reprimands and other adverse actions, we strongly recommend that you contact your local Bar Association before pursuing any specific course of action. It may be that the campaign you've planned is acceptable or you may find it needs a few tweaks to meet the regulations set out by your Bar.

In either case, you won't know unless you ask.

Now, with the legalese out of the way and assuming that you're well-versed on any limits or restrictions imposed by your Bar, here's how to get the most from this book:

We've structured the content contained herein so that even the most novice of users can come away with a clear understanding of how online marketing works and how it can be used to promote your law firm. That means that we started with the most basic of concepts as a foundation and worked outward from there.

Of course, you might not be the most novice user. Quite the contrary, you might already be very familiar with blogging or perhaps be a power-tweeter in your spare time. As a result, you might be tempted to skip certain parts of this book, believing that you already have the skills or knowledge necessary to incorporate that particular component into your marketing strategy. But, we're going to suggest another approach: If you want to get the most benefit from this book, read it in its entirety and in the order it's laid out. And here's why:

Online marketing is a complex subject and one that is constantly changing. So, while you might already know about Facebook or article marketing, you may not know ALL there is to know about those two topics. Additionally, what you do know may have changed recently and the new tools or methods presented here could drastically simplify your campaign.

But we're not done yet! Here's the other half of your success equation:

Because online marketing is constantly changing, it's conceivable that some of what we've written here will no longer apply in the near future. Case in point: just as we were getting ready to go to print, Facebook changed the way they allow users to create customized fan pages, making the instructions we had included here completely useless. Back to the editing room we went.

So, while we've gone to great lengths to include as much up-to-date information as possible about a variety of strategies, tools, and methods, we're also keenly aware that many of those strategies, tools, and methods can change without warning.

That means you'll need to stay on top of new marketing trends, even as you're learning the ones we've outlined here. A good place to start, of course, is our blog—**bestlegalpractices.com**—but don't stop there. Being a savvy Internet marketer (okay, Internet marketer/attorney) means immersing yourself in the trends and technology that make up that marketplace.

Now having said all that, we have no illusions that you'll read this book from cover to cover. We're aware that many of you will probably hand it off to your assistant, admin, or paralegal and ask him/her to report back with a plan of action. Just make sure that plan of action is an informed one, based on the knowledge that you (or your assistant) gained from studying this book and learning about all the variables that make up online marketing.

If, however, you find that you don't have the time, staff, interest, or desire to become the online marketing expert you need to be, there is another option: Skip to the end of this book, page 178 to be exact, and find out how you can go back to practicing law while someone else (that would be us) does all the work for you.

Other Books by the Authors

The E-Myth Attorney: Why Most Legal Practices
Don't Work and What to Do About It

Protecting Your Pet's Future: The Definitive Guide
to Planning When Your Pet Outlives You

Estate Planning Basics: A Crash Course in Safeguarding Your Legacy

Total Wealth Management: The Definitive Guide
to Estate and Financial Planning

Creating a Loving Trust Practice

Part 1:

Gearing Up For Change

Chapter 1:
That Was *Then,* This Is *Now*

Once upon a time, there was a clear and singular path to becoming an attorney. You attended law school, passed the bar, and then joined an established and reputable firm.

Taking this path had clear advantages: You didn't need any real business skills because your success depended solely on your lawyerly talents. Going to work for an established firm meant that all you had to do was practice law. Someone else handled all that business stuff, such as: billing, personnel, and marketing. If you could win a case, argue a point, and/or counsel your clients effectively, you could count on having more than enough work to keep you busy.

And there was certainly plenty of work to go around.

Back then, being a lawyer meant you were in a very exclusive club—one that offered both security and longevity—because, when it came to matters of a legal nature, there was simply no substitute for our services.

This expertise brought with it great authority, allowing us to call the shots and charge substantial fees to do so. It didn't matter if our firm wasn't *"client-oriented."* It didn't even matter if we lacked the skills to run a business properly. As long as we excelled at winning those cases, arguing those points, and counseling those clients, our path remained clear and the money continued to flow.

Of course, this arrangement was not without its flaws.

To walk this path, we had to be willing to give up certain things. Work had to come first because that was the only way to keep those big bucks coming through the door. We were expected to sacrifice our families, our personal endeavors, and even our own happiness, if the firm required it. And it usually did. Time was money. Right down to the last billable hour.

But then something changed.

Enough, we said. *We want something more.* So, we ventured out on our own, determined to create the practice we had dreamed of way back when.

Unfortunately, we still lacked the knowledge to run a business, a fact we could no longer ignore because the buck now stopped with us. We quickly realized that running our own firm wasn't exactly what we had thought it would be.

To compensate for this lack of business savvy, we reverted to doing what we do best—practicing law—and we were willing to do it 24/7 if it meant keeping our firm afloat.

Suddenly, the dream practice we envisioned had, somehow, become the old-style firm we had left behind, just without the safety net we had enjoyed in the past.

But our industry was still changing.

So, while we were busy honing our skills—that is what lawyers do, after all!—the rest of the world was evolving into something we'd never seen before: a digital era that was socially driven.

First, came the emails and then the websites: a static but digital version of our marketing brochure that people could access from their home computers. Of course, this was just beginning. Before we knew it, those home computers seemed to dictate every aspect of *"doing business"* and companies from every industry were scrambling to keep up.

Not us, of course! We didn't have to conform to the regular business world because we didn't own a regular business. *We owned a law firm…* and everyone knows, law firms don't have to abide by traditional business rules.

So, we resisted. We refused to use the email system and we balked at the idea of having our own website. We didn't need it—our clients would always be there.

What other choice did they have?

But the world wasn't through changing and by the time we realized that the digital era was here to stay, we were seriously behind the times.

We tried to compensate by adopting email practices and creating that unnecessary website but by then, our efforts had little impact because our clients had found new ways to get the legal assistance they needed. Companies were springing up everywhere offering our clients the ability to bypass lawyers and their exorbitant fees.

Now, those in need of a divorce or a Will or a Power of Attorney could simply download a form online. And, if there wasn't a ready-made form, there were freelance contractors ready to assist.

Even those clients that still required an attorney's assistance weren't calling as much anymore because, incredulously, they had stopped using the yellow pages and other traditional marketing sources. Instead, our clients were turning to the Internet for referrals. So if you weren't one of the few law firms that had claimed your spot on the World Wide Web, you were essentially invisible.

What's more, our clients had gotten smarter. In the old days, clients had no real way of knowing the value of our services or how we compared to the guy down the street. Now, there was a veritable banquet of this kind of information available. Our clients could access it with the click of a mouse. All the details of who offered what, how much they charged, their experience and what other clients thought—in other words, everything a client need to make an informed decision—was readily available for anyone who wanted to see it.

Oh my! The tables had turned. It was a client's market.

And that's when it hit us: *Our clients had found a work-around. They didn't need us anymore.* And just like that, the days of the old attorney-client relationship were over.

Chapter 2:
The Microchip Revolution

Now, you can blame all these changes on the microchip, going all the way back to its invention in 1959. Of course, society had no idea, at the time, that their world was about to change so drastically. But, change it did! And there's no going back.

Great things can now be accomplished by fewer people and with little or no capital, making today's business world an entrepreneur's market. And what's more, no industry is exempt. Every business is vulnerable.

Yes, even the practice of law.

The good news is that we can look to other industries for guidance. One of the benefits of being the last hold-out is that we can observe what other professions have done and learn from their mistakes.

Accountants are a good example, as are insurance agencies. Even the medical profession has gotten on board with the digital movement, however reluctant some of these moves may have been.

The result is that we have plenty of data to use as a model. We can look at what worked and what didn't and then modify those results to guide us into the future.

But we have to start moving now.

The microchip revolution was really just the beginning. This digital revolution we're experiencing now will continue for at least another ten years. In fact, experts estimate that it won't be fully complete until 2025, and who knows what new changes we'll be facing by then?

Add to that the undeniable move toward globalization and our path becomes clear: either we adapt now or we'll be too far behind to do so in the future.

But, that means you're going to have to change the way you view your law practice. One of the things we teach at Best Legal Practices (and at the Academy as well) is that mindset is everything. You can't conceive a new way of doing business if you're holding onto all the old beliefs and habits that you learned in law school.

Yes, the practice of law is still one that's held in high regard and certainly a profession that will continue to be needed. But, how that profession will evolve depends on how well attorneys can let go of their outdated beliefs.

Your clients won't always be there and your law firm's survival is no longer guaranteed. Getting clients through the door now requires more than just hanging out that shingle and if you want to keep those clients, you'll need to create a business model that is client-focused.

We, as attorneys, have to learn to be socially-adept communicators, skilled in the art of building relationships and masters in things like marketing and strategic planning.

Long story short? We're going to have to evolve.

Chapter 3:
Why Traditional Marketing Won't Work

Now, before we hint at what kind of evolution you can expect, let's talk about why traditional marketing won't work.

It's not because the old, offline methods are dead. Quite the contrary! Many traditional marketing methods are still actually quite effective at building and keeping a client base. The problem is that most attorneys never used these methods to start with because, well… we were attorneys. *We didn't have to advertise.*

Remember, for these attorneys, marketing meant hanging out that shingle and claiming their spot in the yellow pages, playing golf, and going out to lunch—none of which required any real forethought or planning.

Chances are, you're either one of those attorneys or you know someone who is. It's not unusual to find law firms that don't really know how to market their services because, as we've already mentioned, that wasn't a required course in law school. In fact, until to the 1976 landmark Supreme Court case, Bates v. State Bar of Arizona, attorney marketing wasn't even allowed.

The result is that attorneys have spent their careers focused on the technical side of things. We've learned to be the best attorneys we can possibly be within our chosen area of practice without ever having to consider the fundamental mechanics that actually make that practice run.

Sure, we've taken a few courses on time management and, perhaps, even learned to keep our email inbox manageable. But, we've never gotten serious about the business of doing business because we assumed that our law firm was exempt from such concerns.

This assumption, in fact, is what we focused on when we wrote *The E-Myth Attorney*, because it's proven to be our biggest obstacle as an industry. *"We're attorneys, for crying out loud! We're here to practice law. Let someone else worry about business management."*

And that's exactly why traditional law firm marketing won't work: *We never had any real strategies to begin with.*

Of course, we know better now—at least some of us do—and this realization has put us in a very peculiar spot: Do we scrap what we know about running a law firm and start from the beginning or do we just stay on the path we're on and hope for the best?

Sadly, there will be a number of law firms that choose the latter. We guarantee that you'll see those law firms disappear in time.

We no longer have the luxury of resisting new technology. It's this new technology that is now making the rules.

Our clients have embraced it and if we don't get on board, we will most surely be left behind… which brings us back to our original statement—

It's time for attorneys to evolve.

Chapter 4:
The Evolution of the Attorney

We've talked, at length, about where we came from and the old paradigm of practicing law. Now let's look to the future.

Attorneys who adopt the principles we lay out in this book will be indispensible members of our community. We'll be involved in everything from local politics to volunteering, to education—not because it makes us money but because we find such non-firm activities rewarding. For the first time in our lives, *we'll actually have the time to enjoy them.*

Yes, it's true. Savvy attorneys of the future won't be married to their firms because, quite simply, this type of work arrangement will no longer be required. Instead, attorneys will be able to balance their professional lives with their personal endeavors and create an existence that's not just livable—it's actually fun.

This new breed of attorney will still be quite the litigator and/or continue to provide unmatched expertise in his/her area of practice. But we'll also find that these evolved attorneys are skilled business managers as well, fully versed in matters of finance, human resources, strategic planning and, yes, marketing.

Ironically, it will be all this newfound knowledge that allows attorneys the independence they've been seeking all along. Looking back, we'll realize it was the technology we resisted that finally set us free.

We'll likely see more virtual offices and, certainly, virtual employees as law firms begin to recognize the value of flexible work arrangements. As a result, we'll also see more productivity and employee loyalty than we've ever known before.

Client services will still be provided face-to-face where needed but we're betting that we'll find that much of our business can be done virtually and online. Digital discovery and video chatting will be commonplace but we'll also see a new, more personal relationship develop between attorneys and their clients.

This new relationship will be built upon new billing models that give clients a considerable amount of say in how much money they spend on a particular service. The old-time billing models that we know now will be all but obsolete as firms move toward bundling services whenever possible.

In addition, the old rules that govern attorney marketing will become extinct. This is something we're already beginning to see. As a result, law firms will become quite skilled at launching marketing campaigns to promote those bundled services and, subsequently, dominating their marketplace.

The attorney, as we know him/her today, will have evolved into a new, business-savvy entrepreneur, assuming, of course, she/he pays attention to the signs we're seeing today.

Now, the real question becomes: Are you ready to join this revolution?

Chapter 5:
Why You Need an Online Presence

Since you're still here, we're going to assume that your answer was yes. You're ready to become part of the digital revolution that's sweeping the world.

Now, to do that, there are several beliefs and practices you're going to have to modify. One of them is your understanding of, and competency in, online marketing. Of course, that's what you're going to learn here.

So, the first thing we need to do is get past this idea that an online presence is not important. As you've already seen, it will, ultimately, become the cornerstone of your marketing campaigns.

Having established that fact, let's talk about what an online presence actually is.

Dominating the online marketplace does not simply mean that you have a website. Incidentally, if you think that that outdated, static set of web pages you created a few years ago counts as a website, you're sadly mistaken.

An online presence means that you've positioned yourself in your particular market so that you, and your firm, are seen as the preeminent resource on your given topic. In the world of books, for example, Amazon is king. If you decide you'd like to purchase something to read, where's the first place you'll go?

No one has to suggest you try Amazon. You already know it's there. That is, you, and the fifty gazillion other people that shop at Amazon daily. Likewise, if you want to look up something academic, what website pops into your mind first? Wikipedia, perhaps? Once again, no one had to suggest this site. We all just know that it's a good resource to have.

The same needs to be true with your law firm. Maybe not on a worldwide scale but certainly within your community and ideally, within your state. Your firm needs to be the go-to expert on matters of divorce or bankruptcy or you-fill-in-the-blank law and to do this, you'll need to utilize a variety of different tools. Your website is one of course, but, you'll also need a blog, active profiles on social media platforms, and a variety of downloads, freebies and interactive communication tools.

Are your eyes glazing over yet?

Well, hang on, because we're not done. In addition to blogging, newsletters, giveaways, and social media, you'll want to participate regularly in article marketing, backlinking and video creation, plus a variety of bookmarking, ad placement strategies, directory submissions, and the like. And let's not forget the mobile revolution that's taking place. If you're planning to be a leader in your particular area of practice, you'll want to master mobile marketing too.

Now, if you're thinking that this sounds like a lot of work, you'd be right. It is. But then, we've got a lot of catching up to do, remember?

The good news is that you've got everything you need to develop a successful online marketing campaign right here in this book. We've already done the research, tested the techniques, and formulated the strategies for you. Even if you decide to hire a professional firm to do this work for you, you'll be armed with enough knowledge to choose that firm wisely. Read it, live it, and let this be the first step in your evolution.

Now then, let's learn how to market online.

Part 2:

On Your Mark

Chapter 6:
What Is "Online Marketing"?

Wikipedia defines online marketing as the *"marketing of products or services over the Internet."* But, it's actually evolved into much more than that.

When the web first started to become popular, marketers began looking for ways to take advantage of the captive audience it offered. In the past, we had relied solely on offline marketing methods: networking, television and radio ads, newspapers, yellow pages, direct mail and, of course, educational events.

The online venue, however, presented some new alternatives for reaching potential clients including: email, ezines, and pay-per-click advertising. These alternatives were often more cost-effective than many of their offline counterparts and certainly made marketing a little easier. But, marketers still faced the same problem: How could they ensure their target market was receiving their message? You could send an email blast to your mega-opt-in list, just as you could place an ad in the yellow pages. But there was no way for you to reach out to your prospective client when it mattered most: when they were actually thinking about filing for bankruptcy, starting a new company, getting a divorce, creating an estate plan, or even looking for a good litigator.

The answer, of course, was to do more marketing. Like the traditional offline methods, this version of online marketing required that you play a numbers game. To increase your sales, you had to contact more people, buy more ads, and issue more press releases. This made online marketing a tedious and potentially costly venture.

Fortunately, the industry was still changing and, as you're going to see, it was the consumer that finally pointed marketers in the right direction.

Inbound vs. Outbound: What's the Difference?

The problem with marketing up to this point was that it was intrusive. It required you to reach out constantly to the masses and hope that a few receptive prospects might be hidden in the mix. Because your success relied solely on the amount of marketing you could do, the intrusive often became the obnoxious as companies felt the need to bombard the public with their advertising messages.

What's more, many of these messages went one step farther by requiring the prospect to take some sort of big action. So, not only were you taking a gamble that someone might be interested in your seminar, you were also assuming that they would be willing to take off work, skip dinner with their family, and drive across town to your hotel of choice to attend.

Marketing mogul Seth Godin dubbed this typed of marketing as **interruption marketing** because the tactics can get the potential consumer's attention only by interrupting his/her normal daily routine.

It wasn't long, of course, before consumers started looking for a way to block this intrusion. With the help of tools such as caller ID, spam filters in our email, and DVRs to skip the commercials, we're now able to weed out many of the marketing messages before they reach us. Unfortunately, your marketing messages may find their way into that junk pile—not because your service isn't beneficial but because the prospect wasn't expecting or necessarily wanting your message.

It was this concept that prompted the CAN-SPAM Act of 2003, a law that imposed limitations on how and when companies could send out those marketing emails by requiring a relationship to exist before an email could be sent. *Wait a minute! A relationship?*

That gave marketers an idea! What if there was a way to get the client to request the contact?

And this is how **inbound marketing** came about. The goal of inbound marketing is to bring your audience to you instead of the other way around. It is attraction-based, rather than intrusion-based, and focuses more on creating and nurturing the company-client relationship than pushing your wares.

Utilizing both online and offline marketing tools, inbound marketing centers around creating a hub of information and resources that will appeal to your particular market when they need your services. You'll still offer your newsletters, seminars, and podcasts to the public, but the people subscribing and attending are already looking for products, services, and information in your specific niche. This means they're already receptive to what you have to offer.

It's often been said that marketing is like a moving parade. People are fine one minute and then need help the next. No one thinks about new tires until one blows out. Estate planning is a distant thought until a parent dies and reminds you of your mortality.

As Mike Volpe, Inbound Marketing VP for HubSpot puts it, *"Inbound marketing is doing all the right things so that when people are out there on the web, they can't help but bump into you—almost by accident."*

And bump into you they do. To date, inbound marketing has been proven to be the most effective way to market your services and what's more, it costs less per campaign as well.

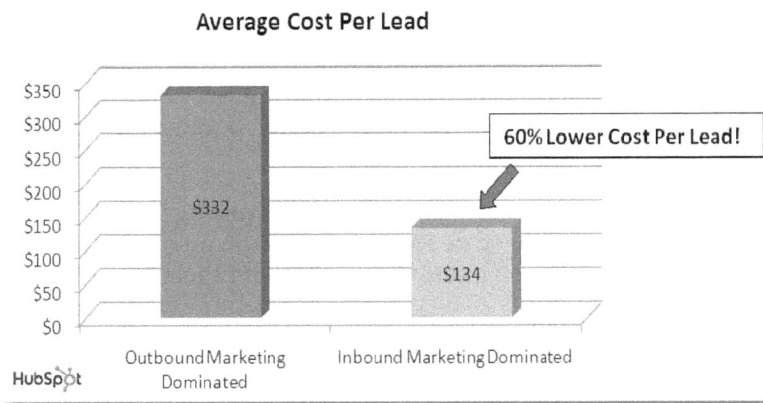

Now, to make this concept work, you have to have a way to deliver your message to the Internet so that it can be found by your target market when they need you.

---

1   HubSpot—*The State of Inbound Marketing Report 2010*—www.hubspot.com/Portals/53/docs/resellers/reports/state_of_inbound_marketing.pdf

Search engines presented marketers with the first solution to this problem. As they gained popularity among web users, marketers realized there was enormous power in where your site ranked in the search engine results.

It was this mindset that spawned the concept of search engine optimization (SEO), a process that allowed webmasters to make their websites more search engine friendly and, thus, easier for their prospective clients to find.

And SEO continues to be a big part of online marketing today.

As we write this, it's estimated that 89% of the population consider search engines to be their primary resource for finding information. They'll log on before turning to other resources such as: the yellow pages, encyclopedias, dictionaries, the phone book, and the like.

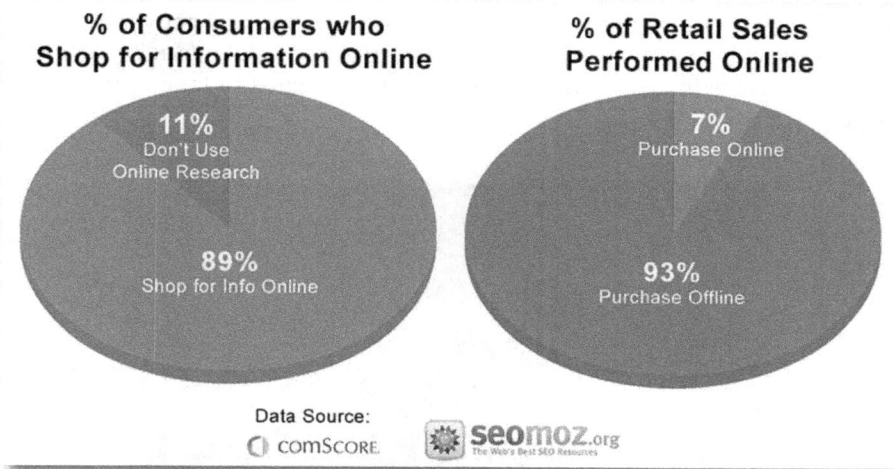

% of Consumers who Shop for Information Online

11% Don't Use Online Research

89% Shop for Info Online

% of Retail Sales Performed Online

7% Purchase Online

93% Purchase Offline

Data Source: comSCORE seomoz.org The Web's Best SEO Resources

The word *"attorney"* for instance, is searched an average of 204,000 times every day (6,120,000 / 30) in the United States, according to Google's keyword tool.

Keyword	Competition	Global Monthly Searches	Local Monthly Searches	Local Search Trends
attorney		7,480,000	6,120,000	

"Bankruptcy" generates about 91,000 searches a day and *"divorce"* comes in at just under 14,000 searches per day. Clearly, people are searching for your services. Needless to say, the competition to be listed in the search engines for terms like this is quite high.

How high?

When those 6.1 million people search for *"attorney,"* for example, Google generates well over 100 million results. This means that there are over 100 million other entries from law firms, businesses, and blogs that are marketing to the same audience as you are.

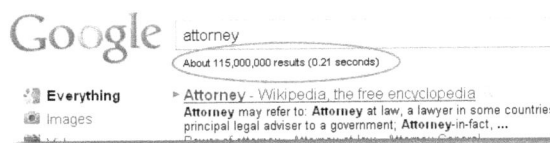

Unfortunately, those 89% who are relying on these search engines can typically find what they're looking for on the first few pages of their search results—often even on page one. So it's rare that a user will click beyond that first page, and certainly not beyond pages two and three.

In fact, marketing firms Enquiro and Did-it partnered up with the eye-tracking firm Eyetools to determine exactly where users looked first. The result was that most of the eye tracking activity occurred at the top left of the search engine results page, forming what is now referred to as *"Google's Golden Triangle."* This triangle reveals that the most activity takes place in the top, left corner and then begins to fade as users look to the right and farther down the page.

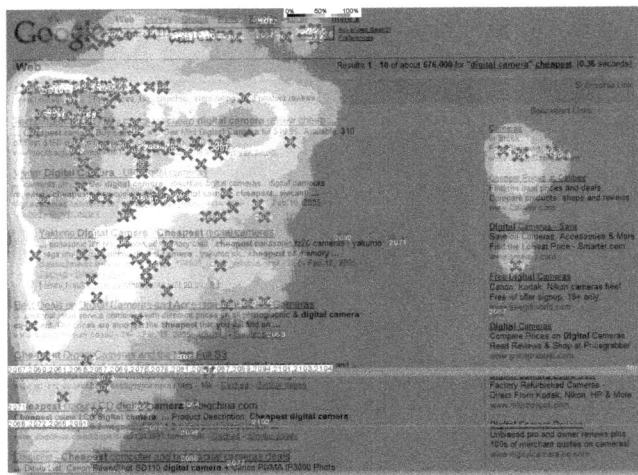

The last listing shown *"above the fold"* averaged 85% visibility, while those below the fold dropped drastically, ranging from 50% to 20%,

So, how does all this eye-tracking activity affect your website?

In a 2006 AOL study, researchers found that 89.82% of all non-branded (generic) search engine clicks came from page one while page two received only 10.18% of the clicks. What's more, the number one listing received just under 50% of all those first page clicks.

Click-Through Ratios for Page 1 SERP

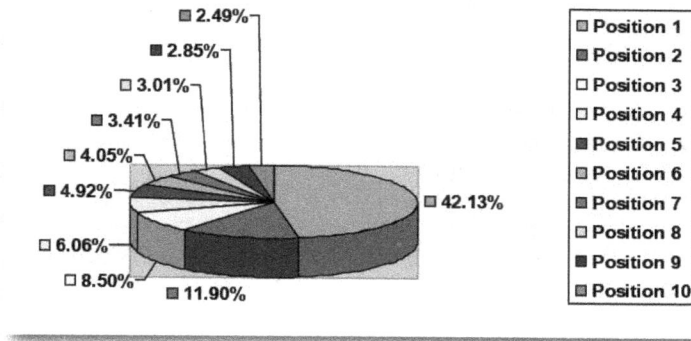

- 2.49% — Position 1
- 2.85% — Position 2
- 3.01% — Position 3
- 3.41% — Position 4
- 4.05% — Position 5
- 4.92% — Position 6
- 6.06% — Position 7
- 8.50% — Position 8
- 11.90% — Position 9
- 42.13% — Position 10

That means, to benefit from being listed in the search engines, your listing should be on page one, preferably above the fold (top five listings) if you want to see any big results.

But with just ten listings per page, getting to page one becomes a tall order—especially when you consider how many of your competitors are vying for that same spot. With so many websites to choose from, the search engines had to come up with a formula to determine their rankings. These formulas are somewhat secretive and extremely complex. But, to put it in the most basic terms, the more popular and relevant your website is to the search terms that are entered, the higher up in the search engines you'll climb.

Now, optimizing your site is a good step in the right direction. But, SEO alone won't give you the market domination you're looking for, contrary to popular belief. Instead, you'll need to add two more components to your online marketing arsenal, components that, when combined with SEO, make up a three-pronged approach to being *"everywhere"* without intruding on your target market and thus, complete the inbound marketing process.

Yes, we're talking about social media and content creation.

Keywords and meta-tags, for example, can help make your website more *"search-engine friendly."* However, social media and blogging will give your online presence a community aspect, something that entices prospective clients to keep coming back.

What you're going to learn here is how to combine these three strategies to create an online presence that *draws the client to you*. We'll start by implementing the basic concepts and then we'll adjust and expand as we go.

Now, that last part is extremely important. One of the biggest mistakes new marketers make is following the *"rules"* and then wondering why they're not seeing any results. The truth is: You won't know what your clients are looking for until you get out there and ask them. That's why a good marketing strategy follows a continuous action-analyze process. In fact, the Internet marketing process can be broken down into three basic components:

1. Research and development
2. Promotion
3. Analysis and adjustment

We're going to explore these components in detail later on. But first, let's take a quick look at what each component entails and how they work together.

Research and Development

Before you can launch a successful marketing program, a little research needs to be done.

In terms of online marketing, that means things like choosing the right keywords, deciding on a domain name, and determining the best ways to reach your target market. Of course, this assumes that you've already identified your target market. If you haven't, then you should start there.

The development piece of this puzzle represents the creative phase and includes creation of your content, marketing materials and free educational reports as well as the actual design of your website and social media profiles.

Promotion

Once you've got all your ducks in a row, you can officially launch your marketing campaign. This can include a variety of strategies such as blogging, social networking, article marketing, videos and directory submissions, many of which are designed to create *"backlinks"* to your website and show the search engines just how popular your site is.

During the promotion phase, you're also still developing new content in the way of blog posts, landing pages, updated marketing materials, free reports, videos and whatever other content you might want to include. Add this to the social networking, bookmarking, article marketing and directory submissions and it's easy to see why this phase is often the busiest of the three.

Analysis and Adjustment

The only way to tell if your efforts are working is to measure your results. Ironically, this is the step most business owners ignore.

Fortunately, we're not *"most business owners"*! We're systems-tracking attorneys and we're going to do this right the first time around.

There are a number of different ways to measure your results. Many of them are free. We're going to talk more about these methods a little later. For right now, just know that you're going to have access to a wide range of data that will tell you not only how many visitors you had to your website, but also where they came from, what links they clicked on and an array of other information you might find useful.

Mine this data and you'll know what services interest your target market the most. You'll know which pieces of content attract the most attention and which referring sites are sending the most traffic your way.

This is good information to have because it allows you to adjust your marketing campaign accordingly, fine tuning it to meet the needs of your target market.

So far, so good?

Don't worry if this all sounds foreign to you right now. By the time we're done, this will all make sense. The important thing to remember is that these components work best when you're using them together. Your success hinges equally on all of them.

You can't develop the right content, for example, if you don't know your target market. The return you see on your promotional efforts will depend on how well you analyze and adjust your campaign.

It's also very likely that you'll be involved in more than one component at a time. So don't treat these steps as individual stages. They're all part of the overall process.

Analyze as you go. Make adjustments where necessary, while you continue to promote, research, and develop. This is the only way to maximize your online marketing efforts.

6.1: *The Problem(s) with Online Marketing*

As we've already mentioned, online marketing is subjective. What works for you may not work for the guy down the street and vice-versa. Your ideal strategy will, in fact, depend upon a number of variables including: your particular area of expertise, your location and, yes, your target market.

In addition, these strategies continue to change, so what works today may not be your best move tomorrow.

The second problem is simply a matter of time, something that we as attorneys value greatly.

Mastering all these strategies and developing a solid course of action takes a considerable amount of time. That means time that we can't devote to a paying client. Add to that the fact that online success doesn't happen overnight, and it's easy to see why so many attorneys give up on online marketing altogether.

Now, if this isn't challenging enough for you, let's throw another obstacle in your way: a lack of information. Ironically, despite the endless supply of strategies and tools available to boost your online marketing, not much has been written specifically for attorneys, most likely because of that general resistance we mentioned earlier. Our hesitation to embrace the online world has essentially nullified the need for data relating to our profession and as a result, there's not much in the way of helpful hints to market a law firm online. See the vicious circle?

So, with limited time and little guidance, attorneys venture into the World Wide Web to discover that not only do they not know what they're doing, but they also don't really have the time to figure it all out.

And this is when one of two things will happen:

Unsure of what to try first, the attorney attempts all the strategies available. But, because no single strategy receives the time and effort it deserves, the attorney sees little, if any, results and, ultimately, gives up.

The second (and more common) possibility is that the attorney becomes overwhelmed from the outset and, having no real sense of direction, either gives up before he ever really gets started or hires an *"online marketing expert"* who promises the moon but doesn't understand the business aspects of running a law practice. Many of these so-called experts don't have the knowledge, training or experience needed to create a successful marketing strategy for a law firm. However, this lack of knowledge and experience won't become apparent until the attorney has spent tens of thousands of dollars on campaigns that go nowhere.

Clearly, there has to be a better way.

And there is. But before we dive into the world of online marketing, let's get a few basics out of the way.

Chapter 7:
Missions, Visions, and Branding: What Does It All Mean?

If you've done any marketing or strategic planning at all, then you're probably aware of the concept of *"branding."* But just in case you're not, branding is the process of creating an association between your firm and the services it provides in the minds of your target market.

Now, we're not going to go into a lengthy discussion on this topic. Branding could easily be a book all on its own. But there are a few points we want to touch on, because your brand, your mission and your vision have a direct effect on your marketing campaigns.

There are a multitude of good branding examples. For instance, what company pops into your mind when you think of fast food? McDonald's™? KFC™? Taco Bell™? If a friend tells you he/she is looking for a good weight loss program, who do you think of first? Jenny Craig™? Weight Watchers™? What if you needed to purchase a new computer? Where's the first place you'd think to try? If you said Best Buy™, you're not alone. This is exactly why branding is so powerful.

Even more interesting is that there may very well be other products and services out there that are better, but because of the positive association created through branding, your favorites will win out every time.

The reason is simple: Branding not only connects the company with the product, it also reinforces the idea of value, community, and trust. Think about it. The commercials we see on television often depict families enjoying the product together or we see the advertising company coming through for us in a pinch. Hertz's™ *"We'll pick you up."* commercials are a great example. Mom's car breaks down on the way to the big soccer game. But Hertz™ is able to get her and her boys to the game on time.

The result is that we feel like we can count on this particular company when it matters most. Their service gives us a sense of security by reinforcing the idea that we can count on Hertz™ to come through we when need them.

And a brand is born.

Now, you'll notice that in each example, the company's tagline defines and conveys its mission to the public. Yet, interestingly, many law firms purposely avoid creating a tagline, even though it's one of the most important components of a good brand.

A tagline, or *"slogan,"* is a good way to reinforce your core marketing message. Take Nike's ™ *"Just do it!"* and Allstate's ™ *"You're in good hands."* Both exemplify the company's mission. Nike™ wants you to get up and move while Allstate™ wants to protect you and your interests.

Historically, however, law firms have viewed taglines as too *"gimmicky"* for a profession of such prestige. As a result, they have shied away from using them. But that thinking has changed over the years as more and more law firms realize the benefits of having a good catch phrase.

Should you decide to add a tagline to your firm's brand, you'll likely want to enlist the help of a professional copywriter as a bad tagline is worse than no tagline at all.

It should be one line, short, and to the point. But, those few words should illuminate the vision you've created for your firm.

For example, an estate planning firm might convey its brand with a tag line developed by the American Academy of Estate Planning Attorneys™ that says, *"Helping our clients preserve their wealth."* Likewise, a bankruptcy firm could use, *"Helping clients rebuild their lives."* A divorce firm might prefer a tag line that demonstrates their dedication to seeing the client through the difficult process of divorce. *"We're with you every step of the way!"* is a good example, as is, *"Helping you start the new chapter of your life."*

Now, in order to develop a strong tagline, you'll need to be clear about the vision that inspired it. So, before you start designing a logo or crafting a slogan, let's talk briefly about the driving force behind it all.

7.1: Do You Have A Vision?

Actually, we already know the answer to that question. It's YES. Yes, you do. Your vision is whatever you imagined when you first contemplated the idea of opening your own firm. It may even be similar to the grand ideas that were swimming around in your head when you first entered law school.

Once upon a time, you had a dream. That dream is the vision for your firm.

Now, it's very possible—even likely—that your vision has changed over time. Maybe you've realized that being a trial attorney isn't really your cup of tea and you'd like to be more *"behind the scenes"* rather than center-stage.

Maybe you'd like to hire a few more associates, expand into other practice areas, or do more pro-bono work. Whatever it is that you'd like to see your firm become, *that* is your vision.

So, how does your vision differ from your mission statement?

Your vision statement describes the *"why."* Why are you in business? What purpose does your firm serve? How does its existence benefit the rest of the world? How will you know when your firm has become what you wanted it to be?

This can be defined by looking at the *"what."* What it is that you want your firm to do? What services do you want it to provide? What people do you want it to employ? What clients do you want to service?

Your mission statement, on the other hand, describes the *"how."* How will you achieve these goals? What actions will you take to create your vision?

In short, your vision looks to the future and sees results. Your mission looks at today and sees the action required.

Now, having defined these concepts, the next question becomes: *Do we really need both?*

And again, the answer is *yes…* yes, you do.

Your vision is your guide. It's what tells you, *"We're going **here**. This is where we want to be."* Your mission statement is your map. It's what says, *"This is the best road to take."* The statements are interdependent. If one changes, the other must follow.

These statements together, of course, provide you with a solid foundation with which to create your firm. It's these statements—or more precisely, the ideas and concepts within—that dictate your branding choices, your hiring decisions, and everything else you do in the name of your firm.

Put simply: If you don't have a mission, and if you don't have a vision, you don't have a brand. And if you don't have a brand, you don't have anything to advertise.

7.2: *Identifying Your Ideal Client*

We've mentioned your *"ideal client"* several times now. So, let's talk briefly about what exactly that means.

Your ideal client is the person, or *type of person*, your firm is seeking to attract. And obviously, the type will differ for each law firm.

So, to determine who it is you're really targeting, you need to ask yourself some questions about the characteristics that make up those ideal clients.

Where do they live? Where do they work? Are they married or single? Where do they shop? What do they buy? What is their annual income? What keeps them up at night?

At the Academy, we teach our members to identify the ideal estate planning client by answering these questions. We call them Bill and Mary. And we build marketing and practice management campaigns with them in mind. We can attest to the success of this exercise. Being able to answer these questions allows you to understand what makes your ideal clients tick. When you know that, you know how to market to them.

7.3: *What Happens When Your Vision Changes?*

Other than death and taxes, change is about the only constant in this Universe. In fact, if things aren't changing, you're in a rut.

And because life is always changing, it's conceivable that your vision and your mission will change as well.

Now, changes in your mission statement don't have to be life-altering. Remember, this is the *"how"* part of the equation, the actions you'll take to reach your ultimate goals.

So, discovering that there's a better way to travel that distance is a good thing. By all means, change your mission when that occurs. At this point, you're still headed in the same general direction. You're just taking a different route to get there.

But when your vision changes—and it very well may—it can have a much bigger impact on you, your staff and the firm itself.

A vision change means you're altering your destination. You're no longer going *here*. Instead, you're going to go *there*. And that can affect the morale in your firm, especially when your staff is still focused on the old vision they know and love.

To remedy this—and avoid the inter-office turmoil that will most certainly follow—you need to make sure that your staff is on board. This can be accomplished by helping your staff *"see"* the new vision you've created.

The best way to do this is to show staff members how they fit within the new puzzle, a point that brings us to our next tip...

How Do You Know When You've Chosen the Right Path?

Easy! Everyone nods in agreement.

A good measure of your vision and mission statement is when your staff can appreciate the goals and purpose as much as you do. When you and your team all seem to be working as one because you're all on the same page, then you know you're on the right path.

Now, this isn't to say that you have to sacrifice your own goals for the benefit of your staff. Quite the contrary! It's your firm. You have the final say. But if your goal is to build a firm that's more than just a place to go to work, then your staff should fit within the big picture.

If they don't, expect to see some turnover in the future. But that's ultimately a good thing, and you should meet it head on. In order to build the firm you want to build, you and your staff must be working toward the same goals.

This brings us to the last section in this chapter: mindset. To launch a successful marketing campaign, you're going to need a new way of thinking.

7.4: Getting the Right Mindset

For most people, the idea of marketing centers around the concept of growing your business and making more money.

And ultimately, this is certainly everyone's goal. But if your efforts consist solely of asking prospects to buy your services, you're going to find that you don't get the response you were hoping for—especially in the online marketplace.

This is because the online venue is a social one, more focused on community and sharing than anything else. Yes, this marketplace likes to spend money and if they see you as a trusted resource, they'll spend quite a bit. But to get to that trusted resource status, you'll have to offer them more than just a way to buy what you're selling.

And for the attorney who isn't marketing-oriented to begin with, this can require a big shift in mindset.

This revolution we're experiencing isn't just about new ways to attract business. It's also about changing the focus of the way we practice. In the past, law firms didn't have to worry too much about the *"client experience"* because the client had few alternatives. But, in this new marketplace, that's no longer the case.

Your online presence—and in fact, your law firm, as well—must be client-oriented. This means that your primary goal should be to meet your clients' needs, not just sell them your services.

And yes, there is a difference.

Having a client-focused law firm requires communication… *regular communication between you and your client.* It means that you value your client's concerns as much as your own opinions and you aren't afraid of a little constructive feedback.

Client-focused law firms offer value-added services such as: newsletters, e-alerts, seminars, and the like that keep their clients updated and informed about what's happening in the law and what's happening in the firm.

Of course, to master this type of communication, you'll need to embrace new technology and rethink the way you handle files, track your billing, and process information.

In short, you need to learn to think like your ideal client: What are they looking for? What makes them tick? How can you connect with your target market? Answer these questions, pinpoint your clients' needs, and then build both your marketing campaigns and the very services they promote to match.

Okay, so we've explained Internet marketing and we've discussed the importance of branding. Now, it's time to roll up your sleeves and start planning your first online marketing campaign. Are you ready?

Part 3:

Get Set...

Chapter 8:
Learning the Basics

Your online marketing campaign is going to have several components. This fact causes many attorneys to run the other way. But on-the-job training is really the best way to learn. So, instead of trying to become an expert in every aspect of marketing before you ever make a move, we're going to cover some basics now and then learn the rest as we go.

So, the first thing we need to do is learn the language.

8.1: A Crash Course in Internet Marketing Lingo

Just as practicing law requires you to know some legalese, online marketing requires you to speak its language. So, let's start there. There's a longer list of definitions in the appendices of this book. But, let's look at some of the most important ones now.

Above the Fold is a design term that refers to the amount of space on a given web page that is above the bottom of the user's screen. In other words, everything the user can see without having to scroll down. Incidentally, this term comes from the newspaper printing industry, where the space on the upper part, above the *"fold"* of the newspaper was considered to be more valuable in terms of advertising than the space below the fold.

Blog is short for *"web-log"* and refers to a type of website, or part of a website that presents content as dated entries displayed in chronological order, usually with the most recent content displayed first. *"Blog"* can also be used as a verb, and refers to the action of *"posting"* new content.

Your **domain name** is a unique identifier used to help users find your website. The domain name actually maps to a numeric IP address. But, it's easier to remember **HomeDepot.com** for example, than it is to remember the corresponding IP address of 92.123.69.160.

Hosting is the short form of *"web hosting service"* and refers to companies who provide server space to users for the purpose of setting up a website.

ISP is an acronym that stands for *"Internet service provider"* and refers to the company or provider of your Internet connection, typically your phone, cable, or satellite company.

Your **IP Address** is a numerical label that is assigned to every device connected to a computer network. This label enables the network to distinguish one device from another. So, your computer could have a different IP address than say, a network printer. In terms of Internet marketing, your IP address is the numerical address for the server where your website is hosted. This address can be shared or dedicated, depending upon the type of hosting you have. IP addresses look something like this: 192.112.61.1.

Meta Tags are found in a web page's coding and are used by the search engines for ranking purposes. The most common tags include the Title tag, the Keyword tag and the Description tag.

When we talk about **navigation**, we're referring to the methods you use to allow your users to move through your website. The most common form of navigation is a menu of hyperlinks.

PPC is an acronym for *"pay-per-click"* and refers to an online advertising model that charges the advertiser each time a user clicks the advertisement shown on a publisher's website.

You're going to hear us talk about **page views**, also abbreviated as **PV.** This refers to the number of times a particular web page was loaded in a browser window. This is the same as *"impressions"* and doesn't measure whether or not the user was unique to your website.

SEM is *"search-engine marketing"* and refers to the methods you use to get your website listed in the search engines.

SEO stands for search engine optimization and describes the process of making your website search-engine friendly by adding keywords, meta-tags and the like. SEO is one of the primary forms of search engine marketing.

Social Media Marketing (SMM) refers to the process of using social media as a marketing tool for your business. **Social Media Optimization (SMO)** is actually related to SEM, but focuses more on generating buzz about your social media profiles rather than your website and search engine placement.

You probably already know this one but just in case, **social media** is a collection of technologies designed to encourage participation, communication and collaboration within a given community.

Your **URL** is your complete website or web page's address, i.e., **www.yourfirm.com** or **www.yourfirm.com/page1.htm**.

Got all that? Don't worry if some of it doesn't make sense yet. When we start using these terms in practical applications, it will all fall into place.

8.2: A Few Thoughts on Domain Names

Back in the section on branding, we talked a little about the importance of choosing the right name for your firm. We also hinted at the notion that your firm name would then dictate your domain name, something that was equally important.

Now, we're going to explain why.

Your domain name is what users type in their browser to get to your web site. **www.yournamehere.com** is an example, with *"yournamehere.com"* being the actual domain name.

For most firms, this is simply a version of their firm's name, so The Law Offices of Smith & Davis becomes **SmithDavisLaw.com,** and that's fine. But let's consider some other possibilities as well.

What if this law firm specialized in divorce or family law? Would **SmithDavisFamilyLaw.com** be better? Perhaps. Or, how about **SmithDavisDivorceAttorneys.com?**

Let's say that the law firm was based in San Diego. Another good choice for the domain name might be **SanDiegoDivorce.com** or **SanDiegoFamilyLaw.com.**

Using keywords that relate to your firm is a great SEO tactic—something we're going to cover a few sections from now. It can boost your search engine rankings exponentially.

Does that mean you can't just use SmithDavisLaw?

Not at all! In fact, you can register more than one domain name and have them all point back to your first choice. So, for example, if our imaginary law firm decided that **SmithDavisLaw.com** was the way to go but wanted to capitalize on the SEO power of **SanDiegoDivorceAttorneys.com,** they could purchase both and have the SanDiegoDivorceAttorneys domain *"point"* and reroute its visitors to the SmithDavisLaw domain.

The benefit of this second domain is that, any time someone searches for the keywords *"san diego divorce attorney,"* this domain has a clear advantage over other firms that have just followed the traditional naming process. When the search engines start ranking sites according to relevance, they're much more likely to pick up on a domain that has all the keywords in its name.

This doesn't guarantee you a number one spot, but it does definitely improve your chances.

Now, having said that, you can see how this strategy might create a seemingly endless list of possible domain names. Fortunately, there are some guidelines you can use—in addition to keyword strategy—to help you choose the right domain name(s) for you.

.Com or .Net? What Extension Should You Choose?

The *".com"* is an extension used to denote what type of organization you have. Both .com and .net are considered to be top-level domain names and can be used for any purpose and by any type of organization. Government websites can use .gov while schools and other learning facilities can use .edu. Domains with the .org extension are typically reserved for organizations—non-profit, professional and otherwise. From there, you'll find a whole host of extensions, i.e., .us, .tv, .biz, .pro. Nearly all of these can be attached to your domain name choice to create a separate domain.

What this means is that **SmithDavisLaw.com** is a separate domain from **SmithDavisLaw.net** and **SmithDavisLaw.biz.** and yes, that means that someone else can purchase the .net version of your domain name and do whatever they choose to do with it. It also means that if you want to corner the market on the name SmithDavisLaw, you would have to buy all these extension variations to do that.

But there's currently over three hundred different extensions to choose from. So, buying all of them isn't really a plausible strategy.

Instead, you want to first shoot for the top level domains: .com and .net. If both are available, it's considered an acceptable and even recommended strategy to purchase both and have the .net version redirect to your .com.

.Pro is a relatively new designation that was created to identify credentialed professionals such as yourself and their related entities. So, while anyone can buy a .com, you must be a licensed professional or business entity to register a .pro domain, making this a possible option for attorneys and their firms. Whether you decide to purchase this extension is up to you. The .pro extension will cost more per year than other top level domains and because it's still new, it's hard to say if this extension will catch on. It certainly doesn't hurt to have it. But, you may prefer to devote your time and funds to other aspects of your website.

The last extension you may want to consider purchasing is .mobi. This is reserved for mobile versions of websites, i.e., sites that are configured and optimized for viewing on mobile devices.

Do you really need to worry about having a mobile version of your website? We say yes, and we're going to show you why and how to get the most from a mobile presence in Chapter 11.

Keep It Short (And Simple)

In theory, your domain name can consist of up to 63 characters, including numbers, letters and hyphens, making it totally possible to purchase the domain **Smith-Davis-Barnes-Russell-Law-Offices.com** with plenty of room to spare, but why would you want to?

In all honesty, few people will be able to remember that domain name, so while it might be accurately representative of your firm, it's not going to do much for you in the way of traffic.

This notion of less is more is nothing new. Law firms have actually been shortening their formal names for several years now, all in the name of marketing. Cooley Godward Kronish LLP legally switched to simply *"Cooley LLP"* in 2010. Howrey Simon Arnold & White moved to Howrey LLP in 2005 and McQuireWoods did it all the way back in 2000.

Shorter names are more memorable and, ultimately, more brand-able. So, while you aren't required to cut those named partners from your letterhead formally, it does make sense to adopt a shorter, more marketable moniker.

The same is true with your domain name. Instead of trying to name every partner in the URL, shoot for something short and sweet and avoid the hyphens if you can. People tend to forget to add them and will likely end up on your competitors site instead, completely by accident.

For example, if you choose the site name **Smith-DavisLaw.com** and there's already a **SmithDavisLaw.com**, where do you suppose the majority of traffic will go? That's right. Your competitor's site! All because you wanted to use a hyphen.

How Much Does It Cost?

Registering a new domain name, assuming you're going with one of the top-level extensions, is relatively cheap, ranging between $10 to $15, depending upon the registrar you use. Note that this pricing is per domain, per year and some of the lower-level extensions will be cheaper.

A .com domain at GoDaddy, a popular seller of domain names, will run you $11.99 a year, plus an additional $9.99 per year if you want to keep your registration details private. A .mobi domain, on the other hand, goes for $6.99.

As a comparison, you can register a .com with NameCheap for $9.98 per year or a .mobi for $7.99 per year. These prices include free WhoisGuard private registration.

A .pro domain will cost you a little more, however, and .pro domains aren't offered by all the various registrars, so you'll have to do a little searching to find one that offers this extension. Network Solutions is a good place to start. They offer .pro domains for around $35 per year.

In addition, there are some companies that offer domains for a little less—around $7 or $8 bucks. But, this is often an introductory offer and your renewal will increase considerably. Just be sure you check the fine print before you sign up.

Also remember that, because this fee is per year, you'll need to renew your domain before it expires. If you don't, your website can no longer be seen and your domain becomes available to the public for purchase, meaning someone can (and often does) come in and snatch it up.

The Relationship Between Domains and SEO

While different suffixes such as .com and .net create completely different domains, the prefix does not. Instead, the prefix denotes a subdomain, something you might want to use if you have a piece of your practice that you'd like to separate or *"distinguish"* from the rest of your site.

For example, if you own **SmithDavisLaw.com**, then you could also create a subdomain for divorce or small businesses or elder law. The thing about subdomains, however, is that they're treated a little differently by the search engines.

Subdomains vs. Folders vs. Separate Sites

In the past, subdomains were considered to be separate websites. So, they didn't share in the overall rankings. This meant that webmasters had to do separate SEO and SEM for subdomains, making them a little trickier to manage. As a result, the rule of thumb was to use folders, such as **SmithDavisLaw.com**/divorce/ instead of divorce.**SmithDavisLaw.com**, unless the content in question warranted a semi-separate website.

Now, for some webmasters, this worked great. All the pages within the site shared in the overall page rank. So, SEO was fairly easy. But the search engines limited the number of results they would show from the same domain, meaning that, in addition to your root domain (**SmithDavisLaw.com**), you could have one other page for your website show up on the same page of rankings for a particular keyword set.

For the more aggressive marketers, this presented a problem. The whole point of SEO and SEM is to dominate the search engines. So, ideally, you'd want to see your pages rank as many times as possible, especially for the broader search terms. To solve this problem, marketers began utilizing subdomains and saw immediate results. Search engines didn't acknowledge the relationship between a domain and its subs, allowing marketers to bypass the *"same-domain"* limitation imposed by the search engines.

Some of that's changing however, as the search engines now recognize the relationship between a website and its various subdomains. As a result, they're changing the way they rank related sites. While the two-page limit isn't strictly imposed on subdomains, the search engines are making it harder for multiple subdomains to rank in the same set of ten results

This change is causing marketers to consider using separate domains instead of subdomains, as we mentioned earlier in this section. The use of separate domains certainly has its benefits: no limit on the number of sites that can rank within a set of ten results. But, there's also the drawback: You'll have to do separate SEO and SEM for each site.

Folders (subdirectories) should still be used for various sub-sections of your site, as long as they are related to the main topic. But, be careful how you structure these. Using subdirectories (**SmithDavisLaw.com**/divorce/page1.html) makes it easier to group your pages and setup navigation, but the deeper your subdirectories go, the less likely it is that the search engine bots will find those lower-level pages. Primary subdirectories—**SmithDavisLaw.com**/divorce/— are fine. But, if you find yourself with multiple levels—**SmithDavisLaw.com**/divorce/children/custody/foreign/—you might want to rethink your structure.

Avoiding Duplicate Content

In light of what you've just read, you might be thinking that having multiple copies of your website is an easy way to improve your rankings. Unfortunately, Google (and most of the other search engines) weed out *"duplicate content"* from their results. This means that if you have an article posted on fifty different domains, only one or two of them will show up in the rankings for the same keyword or key phrase.

Does that mean having secondary, keyword-rich domains is useless? Not at all. To get around this duplicate content limitation, you can do what's known as a 301 redirect. This redirect is a piece of coding that you insert into a file stored in your root directory. The coding is read by your browser (and the search engines) and redirects traffic from one URL to another. And here's how you do it:

If your website is written in PHP, you'll want to place the following code in your **index.php** file:

```
Options +FollowSymLinks
RewriteEngine on
RewriteRule (.*) http://www.yoursite.com/$1 [R=301,L]
```

If the site has an .htaccess file, you can use this method instead:

```
<?
Header( "HTTP/1.1 301 Moved Permanently" );
Header( "Location: http://www.new-url.com" );
?>
```

Essentially, you're redirecting your secondary domains to your primary site. So, if you purchased *"NewYorkDivorceAttorneys.com,"* you can have it automatically redirect your web traffic to your main site at **SmithDavisLaw.com.**

Now, in addition to this coding, you'll also want to make sure that your index page redirects back to your root domain.

Why?

Search engines treat **yoursite.com** and **yoursite.com/index.htm** (or index.php, index.asp, etc.) as two different pages from the same website. That means you're splitting links and thus, dividing your ranking potential for that coveted top ten spot. In other words, your root directory and your index page are competing for page rank.

To eliminate this problem, you should add an additional line to your .htaccess file:

```
redirect 301 /index.htm http://www.domain.com/
```

But we're not done yet. You also want to make sure that both the www version and the non-www version of your domain point to the same place. In other words, if someone types **SmithDavisLaw.com**, they should get the same results as if they typed **www.SmithDavisLaw.com.** And no! It's not always automatic.

To solve this issue, place this line of code in your .htaccess file as well:

```
Options +FollowSymlinks
RewriteEngine on
rewritecond %{http_host} ^yoursite.com [nc]
rewriterule ^(.*)$ http://www.yoursite.com/$1 [r=301,nc]
```

Now, there are other variations of these codes and of course, different solutions for asp pages, java and the like. If these codes don't work for you, remove them and find a better solution. Just do a search for *"301 redirect."*

8.3: *What You Need to Know About Hosting*

Purchasing your domain name simply gives you the right to use that name for the specified period of time. But the name alone won't do you any good. You need to have a place to park it on the Internet if you want it to be seen by the public.

This is where hosting companies come in.

A web hosting company provides you with server space in exchange for a monthly or annual fee. This server is basically a computer that's hooked into a network—in this case, the Internet—and it's what allows you to display your website to the world.

When you tell your registrar to point your domain to a specific server, what you're really doing is assigning your domain name (**SmithDavisLaw.com**) to a specific IP address (such as 192.188.1.1).

Third-party hosting accounts can vary in the amount of access and control you have, as well as in the cost. Choosing the right hosting for you then, will depend upon what you want to spend and what you want to do with it.

So, let's look at the different hosting options you'll encounter.

Types of Hosting Accounts

Shared Servers—This type of server is managed by a third-party company and contains accounts for a variety of customers. What this means is that you and the other customers sharing the server space will not have unique IP addresses. If the server's IP is 192.188.1.1, then that will be the IP address for your site as well as other hosting customers such as Idaho Flower Shop, ABC Plumbing and Ginger Roger's personal blog.

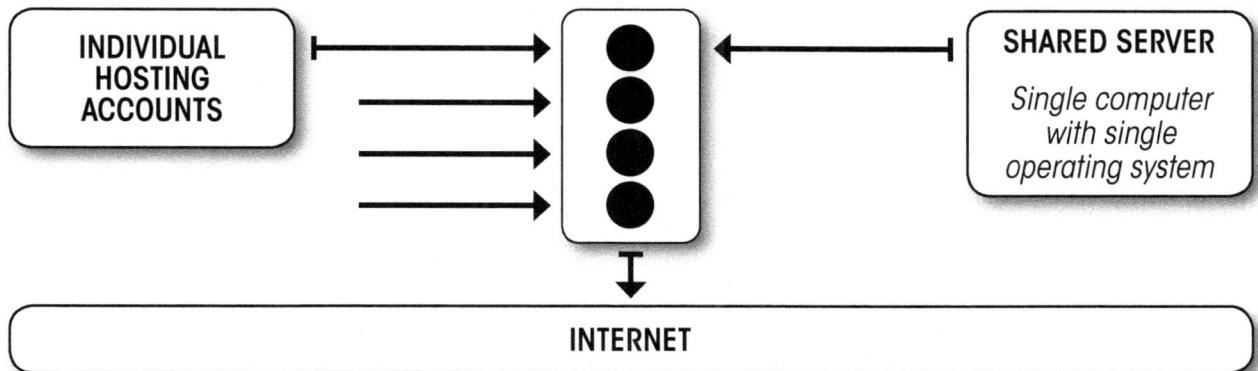

Now in theory, this isn't necessarily a big deal because those accounts won't have any connection to yours (other than their IP address). Your clients won't know that you share server space with a plumbing company or a flower shop in Idaho.

In addition, this type of hosting is relatively cheap—anywhere from $8 to $15 per month. The hosting company is responsible for all the hardware upgrades and maintenance. You'll also find that most companies who offer this type of service have preloaded your hosting account with a variety of tools including a free installer for various CMS platforms, such as WordPress, Joomla and the like.

Your account is then managed by logging into a control panel from the Internet. From here, you can upload files, manage your email accounts and just about anything else you need to do.

But let's say that ABC Plumbing is notorious for spamming. This is a problem with shared hosting because, after a while, the people ABC is emailing will get tired of the spam and complain to their ISP. The ISP will then *"blacklist"* ABC's IP address as a disreputable company. Unfortunately, your website—and any others residing on this same server—will also be blacklisted, because ISPs read the IP address, not the specific domain name.

So, if you try to send email to a potential customer who uses one of those blacklisting ISPs, your email will bounce back as undeliverable. Not so great for your marketing efforts, right?

The good news is that the more reputable hosting companies have strict policies about spamming and take great care to ensure that their IP addresses remain golden. So, your risk here is relatively small.

The other drawback to shared hosting is that you won't have what's known as root access, meaning you won't be able to change server settings or install software that would affect the server as a whole. This, in itself, however, isn't a big deal as most law firm websites won't really need root access.

Virtual Private Servers—Inside your computer, there's a hard drive. On this drive, you've installed your operating system as well as a variety of software programs. It's where you house all your files. When you turn on your computer, this operating system boots up and presents you with an interface to access your files and programs.

But imagine if that one hard drive was divided into sections and on each section was a separate operating system with a separate set of software programs and files. That's what a Virtual Private Server is.

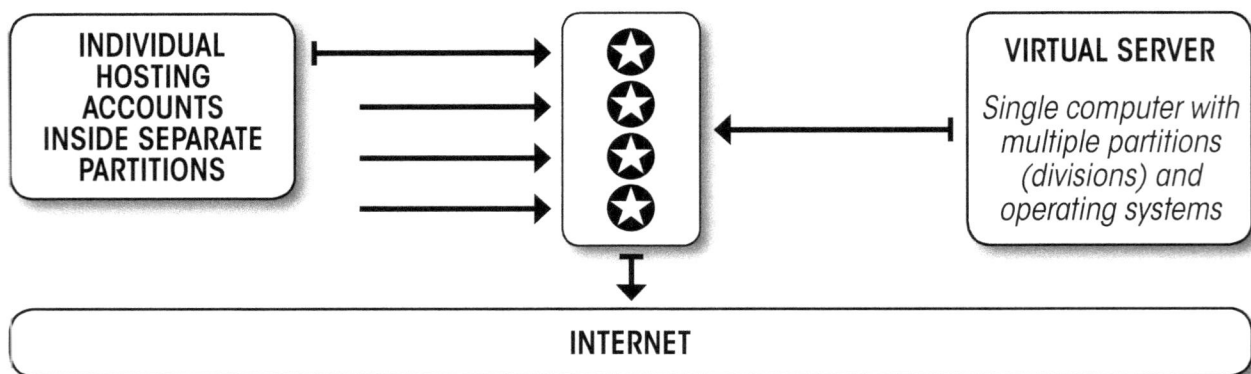

The upside of this type of account is that you have full access to your portion of the server, including that root access we mentioned earlier. You'll still be sharing server space with other hosting customers so you'll still have a shared IP address, but you'll have complete access to your portion of the server, meaning you can install software and make other changes at the operating system level.

Dedicated Servers—A dedicated server is, essentially, a server that is dedicated to your account, meaning that you don't share space with anyone else. This type of server is more expensive for obvious reasons, but it does give you root access and you'll also have your own dedicated IP address.

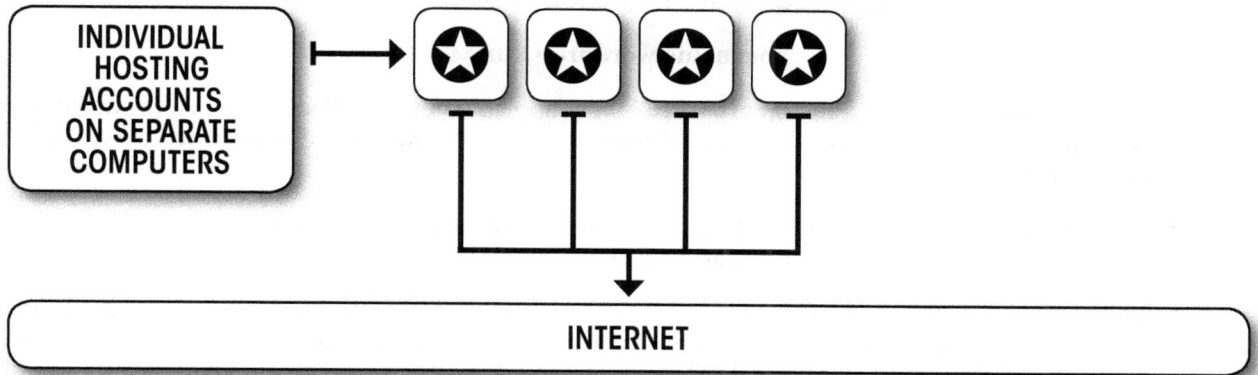

Dedicated servers are a good choice for websites that have enormous amounts of traffic and/or require more bandwidth for other purposes. Again, most websites don't require a dedicated server.

Choosing the Right Hosting Account for You

There are literally hundreds of thousands of companies out there that provide hosting. Many of them are just resellers of bigger companies.

The trick to finding the right hosting account for you is to do a little research and compare. You'll want a company that is reputable and that can offer an account that meets your needs. If your intent, for example, is to host a basic account with a blog, then you don't need an account that accommodates multiple domain names and offers unlimited disk space.

On the other hand, you also want to make sure that you don't quickly outgrow your hosting as paying for additional bandwidth and disk space can get expensive quickly.

In addition, look at *"guaranteed uptime"* as this reflects the hosting company's promise of server performance, i.e., the percentage of time your website is guaranteed to be up and running. Many lower-level providers will guarantee *"97% uptime"* but remember that 97% of 365 days a year equals about 11 days out of the year that you can expect your website to be down for maintenance, server problems or other issues.

While some companies do advertise 100% uptime, it's not realistic. At some point and time, your server will malfunction. It's just the nature of the technological beast. You should, however, be able to count on a pretty-close-to-perfect performance. So look for uptimes that are 99.5% and above.

Uptime stats and other advertised promises aside, read what others had to say about the company. There are plenty of hosting solutions that will meet your needs without making a dent in your bank account. Look around before you buy.

Paid vs. Free

Now, having said all this, there's one type of hosting account we haven't addressed yet, and that's the free account. Believe it or not, you can get web hosting for free from a variety of companies. Some offer it as a way to build their community, such as WordPress.com, while others do it primarily to sell third-party advertising. In either case, your actual URL will be something related to the free hosting company, and not your own.

For example, a free account with **WordPress.com** will give you a URL like this:

yourname.wordpress.com

This is not the same as downloading the open source blogging platform from **WordPress.org** (a subject we're going to cover a little later on). Instead, this blog is hosted on a website owned and maintained by WordPress, and yes, they'll let you have it for free.

The problem, however, is that since you don't own the domain, you have limited control over the amount of customization that can be done and the third-party advertising that might be displayed on your site.

For an individual who's looking to create a personal blog for friends and family, this solution is probably ideal. But, for a law firm, it's not the best option. Instead, opt for a paid hosting account. Many are relatively cheap and, contrary to popular opinion, most law firms don't need a pricey dedicated server to operate a quality website. Quite the contrary, you'll likely do just fine with the cheaper, shared hosting variety. What's more, many shared hosting providers now offer a variety of extra features that include the ability to *"install"* popular open-source software (such as WordPress) to your domain with just a few clicks of your mouse. If you decide that WordPress is the way you want to go for example, just search for hosting providers that offer WordPress as part of their hosting package.

You'll also want a provider that offers a generous number of email addresses (you@yourdomain.com) and one that gives you a reasonable amount of disk space and bandwidth.

HostGator.com is a good one to try. It offers all the features we've mentioned above.

Chapter 9:
Everything You Need to Know About SEO

We've already mentioned that SEO stands for *"search engine optimization."* Now, let's talk about what that means and, more importantly, how you do it.

SEO is a process that ensures your website can be found by the search engines and that it ranks well for certain keywords and key phrases. A better ranking means more traffic for your website and thus, more potential business for you.

Several years ago, SEO was a pretty basic undertaking. All a website owner had to do was plug targeted keywords into a web page, add those keywords to the metatags. Viola! Instant rankings.

Unfortunately, this resulted in quite a bit of garbage showing up in those coveted top ten spots because spammers and other unscrupulous publishers would load up web pages with blocks of unrelated keywords, often *"hidden"* behind the actual content.

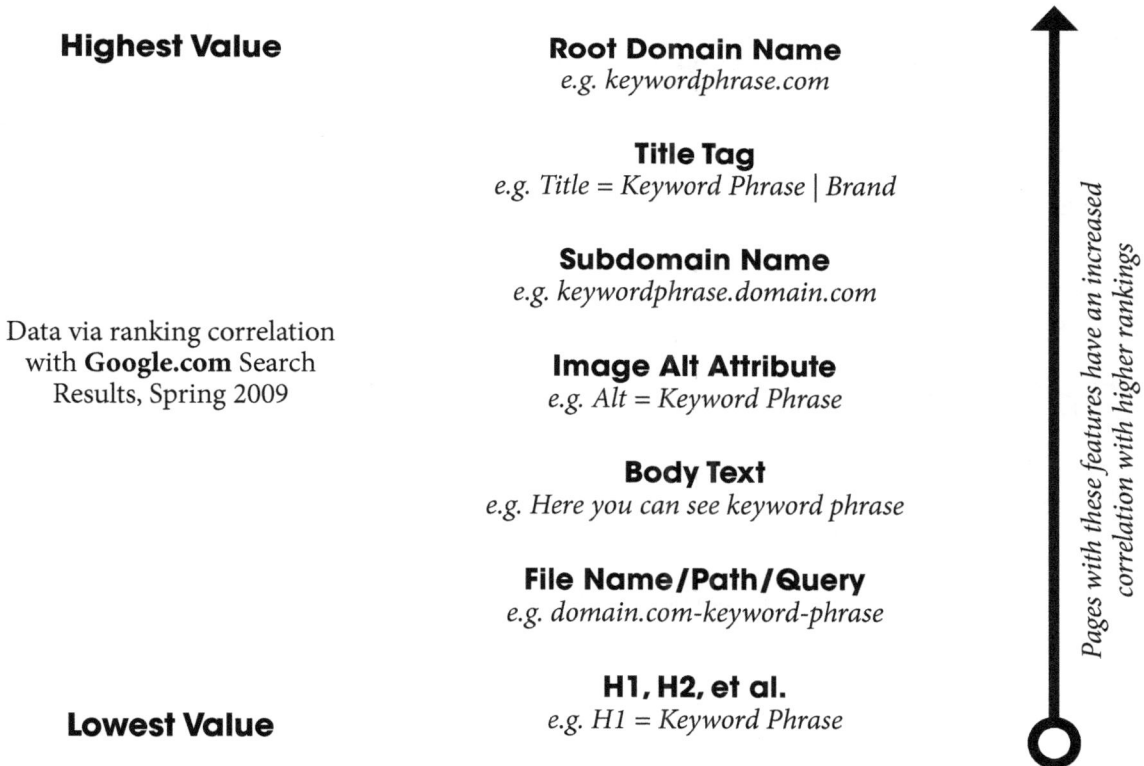

Highest Value

Data via ranking correlation with **Google.com** Search Results, Spring 2009

Lowest Value

Root Domain Name
e.g. keywordphrase.com

Title Tag
e.g. Title = Keyword Phrase | Brand

Subdomain Name
e.g. keywordphrase.domain.com

Image Alt Attribute
e.g. Alt = Keyword Phrase

Body Text
e.g. Here you can see keyword phrase

File Name/Path/Query
e.g. domain.com-keyword-phrase

H1, H2, et al.
e.g. H1 = Keyword Phrase

Pages with these features have an increased correlation with higher rankings

The search engines however, couldn't tell the difference and users ended up at websites that had nothing to do with the information they were looking for. It just contained the keywords they had searched.

To solve this problem, search engines got a little craftier about how they ranked the websites on their list. The result is the complex and secretive algorithms we mentioned earlier.

Instead of just looking at keywords, search engines began to actually measure the value of a site, a move which drastically improved the search engine's results. Using a variety of factors, including the age of the domain name (The longer you've had it the better!) your meta tags, backlinks and keyword density, search engines can actually measure how *"valuable"* a site is to the keywords it's targeting.

All of these measurement tools are found in the search engine's proprietary algorithms and this mathematical formula is so secretive, that only a few of the top engineers in that department know of all the components.

It is, essentially, the secret sauce of each search engine and it's why we rely on search engines so much today. The results they return are, typically, pretty reliable and that's exactly what they want.

A search engine's goal is to provide an easy way for users to find sites that are relevant to their respective search terms. The better the results, the happier the user and happy users will keep coming back. The more users a search engine has, the more revenue they'll make from their paid advertising spots. That's why search engines promote some sites with higher ranking and punish others. They want to ensure their users have a good experience every time they search.

Of course, these complex computations also made it harder for legitimate websites to climb the ranks. That, and the fact that all of your competitors want the same page one rankings you're looking for.

In response, website owners started looking for new ways to boost their rankings and the search engine optimization process was born.

The first thing you need to know about SEO is that it's not a guaranteed ticket to stardom. SEO can do great things. But, SEO all by itself is just one component of a complete Internet marketing campaign.

The second thing you need to know is that, like Internet marketing, SEO is also made up of various components. To truly master search engine optimization, you need to understand how they all work together.

9.1: *The Three Primary Components of SEO*

Search engine optimization can be broken down into three basic components: keyword research, optimizing your site and building backlinks.

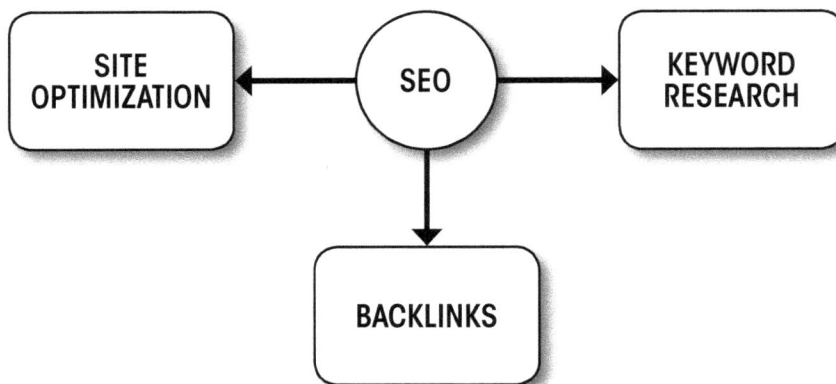

Together, these three components help website owners improve the measured value of their websites, also frequently referred to as **Page Rank**—a key factor in determining how the search engine rank their results on a given search.

Step #1—Keywords

Keywords are words that relate to your particular topic or, in this case, your practice area. In other words, what are your ideal clients searching for? A divorce attorney might choose keywords such as *"divorce, custody, alimony and child support"* for example, while an estate planning attorney might want *"wills, trusts and probate"* as its primary set of keywords.

These one or two-word keyword selections are referred to as **short-tail keywords**.

Keywords can also include complete phrases—known as **key phrases**—to further define the topic. Using the estate planning website, wills trusts and probate could be accompanied by phrases like *"writing your own will, setting up a living trust and new york city probate."*

Longer key phrases are often referred to as **long-tail keywords**.

Now, you can use keywords in a couple of different ways for SEO purposes: the first is by using them in the metatags as we'll demonstrate in the next section and the second is by creating keyword-rich content for your site. This means that if you want your site to rank well on the search term *"wills,"* creating content that talks about wills—such as what it is, how it works and why you need one—is a good way to improve that ranking.

We're going to address content creation later on. But first, let's talk about how to choose the right keywords.

Choosing Your Keywords Carefully
Keyword selection might seem like a no-brainer. But, actually, there's a science to doing it successfully. The reason is that anyone can target the keyword *"estate planning"* or *"tax attorney."* But, how many sites do you suppose can make page one in the search engines with that phrase?

That's right! Ten.

So, while you might want to start with the obvious keywords, a little more research and analysis is required before you have a solid list. And here's how you do it:

Drill-Down
Let's use our favorite firm, Smith & Davis, again. Let's say that this law firm is located in Denver, Colorado with a focus on estate planning.

The most obvious keywords for Smith & Davis then, would be the following:

- Estate planning
- Wills
- Trusts
- Probate
- Denver
- Colorado

But if Smith & Davis try to target the phrase *"estate planning,"* they're going to be competing against national sites such as Wikipedia, FindLaw, and our own American Academy of Estate Planning Attorneys in addition to the multitudes of other law firms across the country that provide estate planning services—and all those online companies who offer do-it-yourself estate planning forms.

Clearly, this is not the best plan.

So, let's dig a little deeper. What kind of variations can we come up with that might fine-tune our keyword list?

For starters, we can combine some of the practice area keywords with the location keywords for a more specific target. So, *"wills"* would become *"Colorado wills"* for example, and *"probate"* could become *"Denver probate"* instead. We could also use variations of these keywords, such as *"Denver probate attorney, Colorado wills trusts, and Denver Colorado estate planning attorney."*

If you were to look at Academy member keyword selections, you'd see phrases such as New York probate, trust attorneys Indianapolis, asset protection planning Denver Colorado and wills trusts Worcester Massachusetts.

What's more, as you delved deeper into the site, you'd find more specific key phrases such as: estate planning dangers San Diego; how to preserve wealth Boulder CO, and what is durable power of attorney Smithtown NY.

These key phrases were selected because they accurately describe the content on the page(s) they represent. So, if someone in Smithtown wants to learn more about a durable power of attorney, our Member is more likely to rank at the top of the search engine's list.

Know Your Clients
The next thing to remember when creating your keyword list is to choose keywords that are actually being searched. Now, using our example above, you're going to find that *"estate planning"* is searched much more than *"Denver Colorado estate planning"* and your instincts might be to go with the words that have a higher search ratio.

But remember, you don't want just anyone looking for information on your practice area. You want potential clients who need your services in your area.

And this is where understanding how your clients search is vital.

Your potential clients might not initially search for an attorney, for example. Instead, they might just search for *"Colorado probate law"* or *"Colorado wills."*

Likewise, many potential clients may not even know what *"estate planning"* is. But they know they need information on setting up a trust. By the same token, someone seeking information about a divorce might not realize that divorce falls under a family law practice. So, using the phrase *"family law"* in your keyword selection might not net you any real results.

Understanding what kind of search terms your target market uses will (or should) directly affect the key phrases you use.

The Benefits of Long-Tail Keywords
Adding your city and state to a keyword helps eliminate national competition. Thinking like your potential customers gives you insight to more common search phrases. But, how do you beat out the firm down the street?

For a new webmaster or Internet marketer, targeting shorter and, thus, broader keywords is often seen as the primary focus. Their reasoning is simple: the marketer or webmaster wants to generate as much new business as possible in the shortest amount of time. At first glance, a broader keyword strategy clearly offers that kind of coverage, ensuring that the website in question will rank for as many related keyword combinations as possible.

New York estate planning attorney, estate planning New York, wills and estate plans… you get the idea.

But if you examine this strategy a little closer, you'll begin to see its flaws.

For starters, someone searching for a broad term key phrase is less likely to have a clear objective when they start their search. Using our example above, the search for *"wills and estate plans"* could be done by someone who actually needs an estate planning attorney's assistance or it could be because the person is looking for do-it-yourself forms. It's also a possibility that the search was conducted purely for research purposes and the user has no intention of buying anything in the near future.

A long-tail keyword search, however, can help refine your traffic so that the visitors to your website are more likely to be looking for you too.

To find these golden, long-tail keywords, you can start with the same process we outlined in the previous section. Choose basic relevant keywords. Then, drill down to add your various modifiers. But, this time, we're going to go one step farther.

When it comes to keyword selection, you have to think outside the box. Start with those common search terms you listed and then get creative. For example, instead of *"Colorado probate law,"* you could use a key phrase such as, *"how to probate a will in Colorado."* Likewise, a divorce attorney will likely see better results by targeting a phrase such as *"do-it-yourself divorce new york"* versus a more generic *"new york divorce attorney."* Yes, these are much longer key phrases. But, they're very specific and tend to generate a better flow of targeted traffic.

How much better?

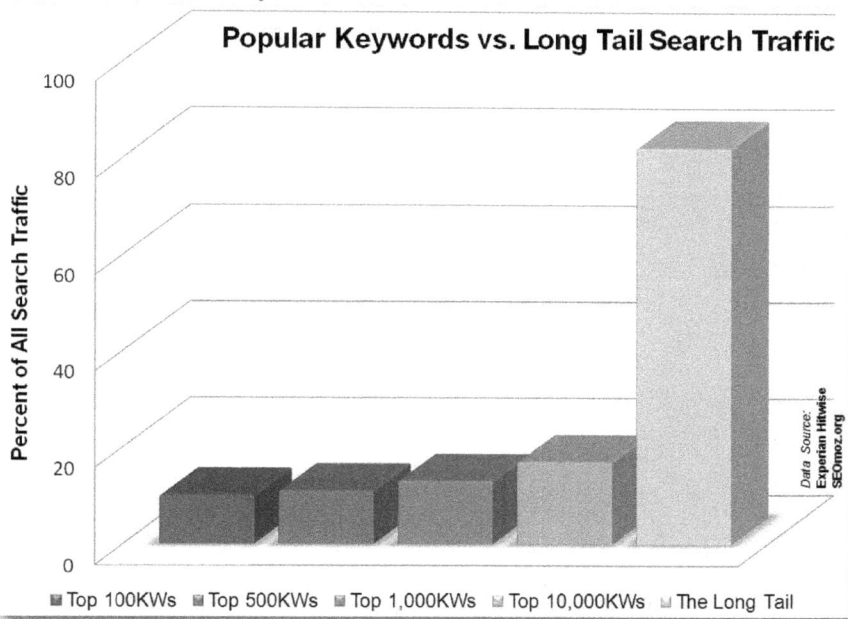

Looking at this chart, you can see that long tail keywords generate as much as 90% of overall search traffic when compared to top (and more competitive) keyword phrases. Amazon, incidentally, makes 57% of their sales from keywords that are considered to be *"outside"* the most popular terms. Given Amazon's amazing success in online marketing, it only makes sense to consider using some of their strategies.

Now, if you're a divorce attorney, you might be wondering why we suggested the do-it-yourself key phrase above when, clearly, you don't sell that service. You do, however, handle divorces and it's reasonable to assume that some of your prospective clients might be considering doing their own divorce and are hoping to find the right guidance from the Internet.

You, of course, know better. So, why not create a good content piece that explains the dangers of doing your own divorce and then using SEO to rank higher for those keywords? Remember, your goal is to rank for the phrases your clients are using in their search and if do-it-yourself (or any other related subject) is ranking well, you need to consider addressing that topic.

Analyzing Your Keyword Choices

You might be wondering how to tell if your keywords are good choices or if you need to go back to the drawing board.

The truth is: Your keyword analysis should continue overtime. With the help of some targeted marketing efforts, you can tweak those keywords and phrases to find the ones that work best for you.

That being said, it would be nice to have some indication, up front, about the quality of your initial choices. To do this, you can use Google's Keyword Tool and roughly measure just how popular your keywords are.

To access the tool, go to **adwords.google.com**/select/keywordtoolexternal. *Note: this works better if you're logged into your Google account.* To use the keyword tool, simply plug in your chosen keyword or phrase in the box, type in the letters shown in the Captcha and click *"Search."* We recommend searching only one keyword or phrase at a time to maximize your results.

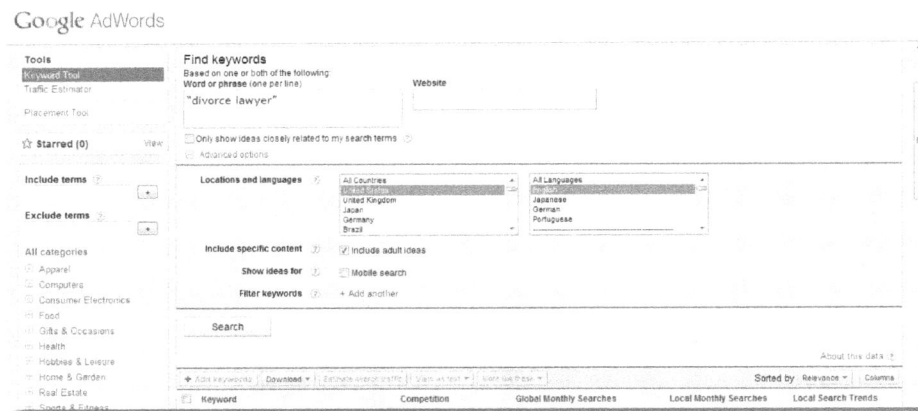

What Google will return is a breakdown of the searches performed on this keyword as well as related variations.

So if we type in the term *"divorce lawyer,"* the keyword tool returns about 800 results.

To analyze these results, you need to understand what the different columns within the keyword tool mean:

Competition—This reflects the number of advertisers that are bidding on this particular keyword in relation to all the keywords up for grabs in Google Adwords. This is displayed as a status bar and if you hover over it with your mouse, Google will tell you if the competition is high, medium or low. If you want to view the actual numbers, you can choose to download this data in spreadsheet format. (**Be sure to check the search terms you want to include.**) You can see the competition in percentage format instead for a more accurate comparison.

Global Monthly Searches—This number represents a twelve-month average for searches on this keyword as recorded by Google.com. Note that this is a **global search**, meaning that the results include users from other countries.

Local Monthly Searches—This reflects the twelve-month average for searches on this keyword from users within your targeted country. If you've logged into your Google or Adwords account, Google uses the information from your profile to filter this category. You can see which countries and/or languages are included in the Local Monthly Searches by looking just below the box where you entered your keyword to search. You'll see a link called *"Advanced Options"* and if there are any location modifiers in use, they'll show up here. If you don't see any modifiers, click the Advanced Options link and add in the United States to filter your Local results.

Local Search Trends—This reflects the last twelve months of search queries so that you can see how a particular search term is performing. If you download your data, this graph will be converted to the actual number of searches per month for better evaluation.

Now, how do you use all this information to filter your keyword choices?

Well, first of all, you'll notice that we started with a very generic term: divorce lawyer. We didn't include any location modifiers because we wanted to see what variables Google returned first.

And at this writing, the term *"divorce lawyer"* is searched an average of 201,000 times per month in the U.S. (based on the Local Monthly Searches). Not too shabby, is it? But just a few listings down, we also see that *"best divorce lawyer," "find a divorce lawyer"* and *"cheap divorce lawyer"* are also terms that are being searched on a regular basis, but with much lower numbers. So, let's look at other highly-searched terms. Click on the column heading *"Local Monthly Searches"* and Google will resort your data in ascending order by the number of average searches performed.

When we do this, we see that *"divorce"* is actually the highest ranking variable, averaging over four million local searches per month. Again, your numbers will probably vary as search terms go up and down in popularity, but you get the idea.

The next most popular search term is *"child support"* with over 2.2 million searches per month. Now, if we wanted to use what was most popular as our only criteria, our work would be done.

But we're not going to do that and here's why:

If you look at the competition heading for these search terms, you'll see that they're both extremely high. That's because every divorce lawyer in the United States (and for that matter, those in other countries as well) are bidding on these two terms. And if we do a quick search on Google for the term *"divorce,"* we're rewarded with a whopping sixty-five million and change in results, the first listing of which is for Wikipedia. In fact, in the first ten listings, not one law firm is shown.

So, where do you suppose your law firm will rank among those sixty-five million listings?

A better idea, then, would be to target keywords that are regularly searched but also offer less competition. To do this, click the *"Competition"* heading twice so that the results are descending and you'll see your collection of search terms sorted by the least amount of competition to the highest.

At this writing we see several options: divorcing, divorce statistics, child support and divorce children all show to have low competition and relatively high monthly search numbers.

So, now that we have a starting point, let's look at how location modifiers can affect our numbers. For starters, we can see that *"attorney new york"* is a potential keyword choice but what if we're not in New York? How do other locations rank?

Well, if we start scrolling through the results, we see that *"California divorce lawyer"* and *"Houston divorce lawyer"* are also being searched (4,400 and 3,600 searches per month respectively) but both have fairly high competition. California child support on the other hand might rank better but this is a lot of listings to read through one by one. Rather than sorting through 800 + results, however, you can either search again and plug in your location modifier or you can click the *"Keyword"* column heading to sort your data alphabetically.

Let's start with the first option. If we add San Diego to our divorce lawyer search term, we get the following results:

Find keywords
Based on one or both of the following:

Word or phrase (one per line)
san diego divorce lawyer

Web site

☐ Only show ideas closely related to my search terms
☐ Advanced options   Locations: United States ×   Languages: English ×

[Search]

About this data

＋ Add keywords | Download ▾ | Estimate search traffic | View as text ▾ | More like these ▾   Sorted by Competition ▾   Columns

Keyword	Competition	Global Monthly Searches	Local Monthly Searches	Local Search Trends
san diego district attorney		3,600	3,600	
san diego county child support		1,300	1,300	
san diego child support services		1,300	1,300	
child support services san diego		1,300	1,300	
child support services in san diego		1,300	1,300	
divorce and		368,000	301,000	

The term *"child support san diego"* looks promising, as does *"san diego child custody."* This search also tells us that the term *"getting a divorce"* has low competition but is averaging about 27,100 local searches per month. If we wanted to capitalize on that particular key phrase, we could create several articles or blog posts and maybe even a free report with those words in the titles. Imagine getting just a portion of those twenty-seven thousand searches each month and each visitor would be looking specifically for information you were providing.

Kind of makes this keyword research all worthwhile, doesn't it?

In addition to Google's free keyword tool, there are also many specialized software programs sold by experienced SEO practitioners that can help you find keywords as well. Which keyword tool you decide to work with is up to you. But, using this same process, go through each of your keyword choices and you'll find plenty of variables to use along the way. Analyze the numbers and you can drill down your keyword list and come up with some powerful choices that generate a steady stream of traffic.

Step #2—Optimizing Your Site

With a solid set of keywords, you can start optimizing your website. This is essentially a two-part process that ensures your website has an abundance of your targeted keywords.

The way we do this is by creating keyword-rich content and then modifying the coding that runs behind your web pages. So, let's start with the content:

Building Content That Drives Traffic

The only reason that anyone is going to want to come to your website is because they believe you have something they need. This *"something"* can range from products you sell, to services you provide, to information you're willing to share.

All of this information will be included on the pages you create for your website and each of these pages should be optimized for your chosen set of keywords.

Starting with your basic pages—Home, About Us, Services, etc.—you'll want to go through the content and make sure it includes your chosen set of keywords.

In general, your **keyword density** (the number of times a given keyword appears compared to the total word count) should range between 3% to 7%. But, that's a target that changes a little from search engine to search engine. As it stands right now, anything over 10% is considered too much and you risk the search treating your content as spam. To put this into perspective, a three-hundred-word article with a 7% keyword density would have the keyword listed twenty-one times.

Now, before you just start randomly inserting keywords into your content, keep in mind that sacrificing quality in favor of keyword quantity is not the best strategy.

If the article is better written with a keyword density of 3%, then go with the 3%. Don't add extra keywords if they don't add to the article.

For example, this screen shot shows an article written about probate from one of our Member attorneys at the Academy:

We work hard to help our clients avoid the time and expense involved in probate. Although there are lots of reasons avoiding probate is a good idea, one major concern regarding probate is that the process can often keep family members from getting access to the property and assets of the deceased in a timely fashion.

Say you are married and have two kids. If you pass away and most of your assets are probate assets, what happens to your family while your estate is tied up in probate? While there may be funds available… your spouse and children won't be able to use them.

The best answer, of course, is to ensure you create an estate plan that helps you avoid probate. An ounce of planning is worth a pound of cure.

If assets are tied up in probate, some states allow surviving spouse and child awards. While the language varies, the award is an amount of money that provides for the support of a surviving spouse and children. The goal is to allow your heirs to access at least some of your estate so they have funds to live on while your estate works its way through probate. Filing a petition with the court handling the probate gets the *"awards"* process started.

Again, the best way to make the best of probate is to avoid the probate process altogether. A comprehensive estate plan will protect your assets, avoid estate taxes where possible, and ensure your family's financial needs are taken care of as soon as possible.

A quick analysis of the keyword density tells us that the most obvious keyword—*"probate"*—has a density level of 2.22%. *"Estate planning"* ranks at 1.04% while *"planning"* has a density of 3.25% and *"estate"* is 2.66%.

Keyword Density

Keyword	Count	Percent	Keyword	Count	Percent	Keyword	Count	Percent
planning	22	3.25%	estate planning	7	1.04%	estate planning attorneys	2	0.30%
estate	18	2.66%	american academy	5	0.74%	business planning elder	2	0.30%
probate	15	2.22%	business planning	3	0.44%	planning elder law	2	0.30%
resources	6	0.89%	retirement planning	3	0.44%	ira retirement planning	2	0.30%
academy	6	0.89%				major concern regarding	1	0.15%

Ideally, we'd like for our density to be a little higher. However, this article reads well the way it was written and it's also rather short. Adding additional keywords just to meet density ratios would reduce the quality of the content and that's a big no-no in the world of SEO.

The other thing you need to remember is that every page does not have to include all your keywords. Quite the contrary, some pages will reflect only one or two keywords and that's fine. Your collection of keywords should be represented by your site as a whole, and not on every single page. Adding keywords to unrelated content is considered bad marketing and could be tagged as spam by the search engines.

Going Beyond Your Basic Pages

Now, we've touched on using your content to target specific keywords and phrases in previous sections and that's what we want to address here. In addition to those standard pages we mentioned earlier, you'll want to add additional optimized content to your site on a regular basis. This serves two purposes: The first is that it gives the search engines more pages to **index** (the process of adding a web page to the search engine database) and, yes, indexed pages contribute to your overall page rank in the search engines. The second reason for creating all this additional content is to give your potential clients something to sink their teeth into.

Remember, we want to attract potential clients that are likely to retain your firm. To do that, we have to get them to your site AND convince them that you're the go-to guy for this particular area of practice. Unfortunately, they're not just going to take your word for it and as far as law firm websites go, they all look pretty much the same to the average web user.

So, the only thing to keep a user from clicking *"back"* and away from your website is to offer them something that makes them want to stay. And in terms of Internet marketing, that *"something"* starts with content.

The content you put on your site can include blog entries (something we address in-depth later on), informational articles, free educational reports, videos or podcasts. Each of these components offers the general public a reason to visit your site. They also give you and your readers plenty to choose from when creating backlinks, the third component of the SEO process.

You can create articles that address specific concerns within your practice area, such as: what to expect in a divorce, the difference between an S Corp and a C Corp, how recent changes in the tax laws will affect small businesses, and what to do if someone is infringing upon your trademark.

What's more, you want to break each topic up into as many different articles as you can. For example, our member attorney, Cheryl David, in Greenboro, North Carolina, covers a number of different topics in her blog:

Simple Tools for Avoiding Probate: The Joint Account

Jan 10, 2011 / By: **Cheryl K. David, Estate Planning Attorney** / Category: Estate Planning, probate

A joint bank account is a simple and very common way to keep a portion of your assets out of probate when you pass away.

When you hold a bank account jointly with one or more other people, and that account is designated as having "rights of survivorship", then the assets in that account pass directly to your remaining account holders at your death.

As simple as this option is, there are a few drawbacks:

1. **The only people to inherit the account are your surviving account holders.** The assets in a joint account won't pass to the beneficiaries you name in your Will. Further, if you intend to add a family member to the account, but never get around to it, that person is out of luck. The money in the account goes directly to those named on the account.
2. **Your co-owners' money problems could become your money problems.** Each co-owner of a joint account have access to all the money in the account, so one co-owner's poor spending decisions can deplete the account assets. Plus, if your co-owner is sued by a creditor, all the assets in your joint account could become subject to collection by that creditor.
3. **A joint owner who is a minor could encounter complications.** When you pass away leaving money to a minor, that money has to be managed on the child's behalf until he or she reaches adulthood. When it comes to a joint account, this could mean that a guardian might have to be appointed to handle the money for your young co-owner. This involves going to court and having a judge make a formal appointment, turning a "simple" probate avoidance method into a complicated process.

These complications aside, joint accounts are not the solution for transferring all of your assets outside of probate. Our team can help you coordinate a variety of methods to transfer your property to your loved ones as efficiently as possible. Call us today – (336) 547-9999.

SUBSCRIBE TO OUR BLOG!

Enter your email address:

Subscribe

Note: Be sure to confirm your subscription! Please check your inbox and click the link provided to confirm.

CATEGORIES

Elder Law

Estate Planning
- Estate Tax
- IRAs
- Probate Questions
- Tax Credit for Home Purchase

Financial Planning

Funeral Arrangements

Incapacity Planning

LGBT

In addition to the article you see here for *"Tools for Avoiding Probate,"* Cheryl also has articles that tell the reader: how to avoid probate with a living trust; how your final bills are paid during the probate process and how a holographic will affects probating your estate.

Likewise, under the category of Incapacity Planning, Cheryl has several articles addressing each aspect of this complicated issue:

- What is a Negative Inheritance?

- Estate Planning Concerns for Same Sex Couples

- What is a Power of Attorney?

- Why a Will is Not Enough?

- Two Essential Documents Your Aging Parents Should Have

- What Makes a Durable POA *"Durable"*?

Creating this hub of content on each topic area does two things. First, it supports Cheryl's assertion that she's THE estate planning attorney you want to have in Greensboro, North Carolina. She clearly knows her stuff and she's dedicated to ensuring that you understand all your options when making these important decisions.

The second thing this strategy does is it tells the search engines that Cheryl is a valuable resource for her selected keywords and phrases. She doesn't just mention estate planning and probate a few times in her website. She's got dozens and dozens of pages dedicated to those exact topics. What's more, she's continuing to add new pages, and that tells Google and the other search engines that she's a good resource to have.

The result is that Cheryl's website ranks well for her chosen keywords. A quick search for Greensboro, NC estate planning for example, returns about 119,000 results, with Cheryl's site listed first.

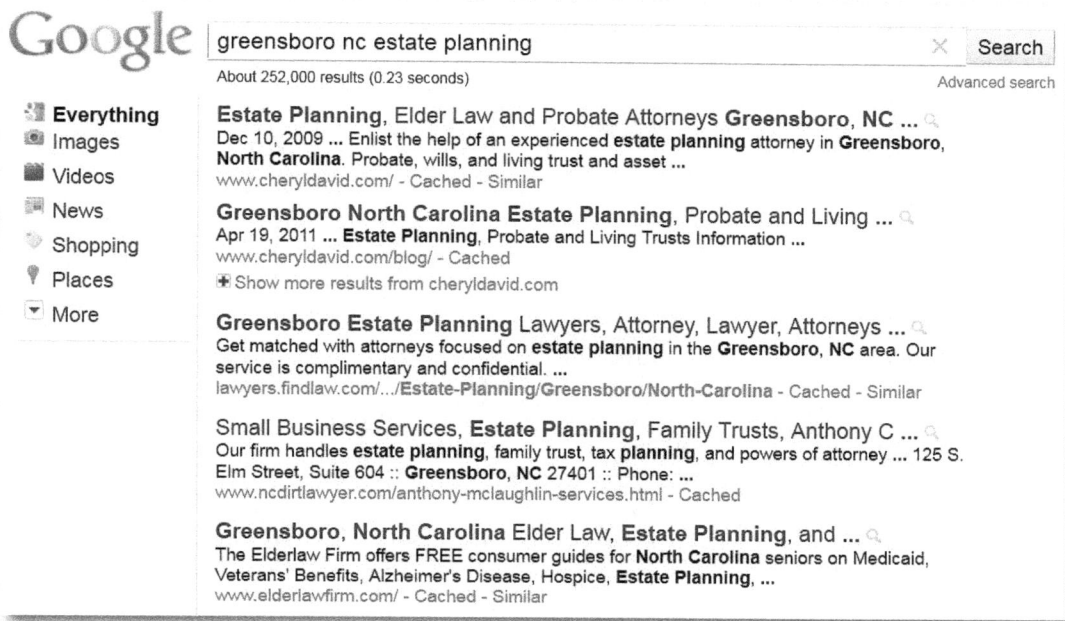

Now, that first listing is Cheryl's Google Places listing, a local search strategy we're going to cover later on. The second listing is the general organic search results from Google's page ranking system. So not only does Cheryl land the first spot in Google, she essentially dominates this search term.

Incidentally, Cheryl also has the same dominant ranking in Yahoo and Bing, too. She is representative of all our Academy members who've relied on us for their online presence.

Which of course, is exactly why we're here.

Going Behind The Scenes—Optimizing Your Code

Every page on your website is made up of coding. The most common code is hyper-text markup language or HTML for short. Your pages can include other types of coding and scripts as well, such as PHP, ASP, Java and the like, but the concept of optimizing your pages is the same—you want to add code to each page that increases its perceived value by the search engines.

And there are a couple of ways to do this.

Metatags—Metatags are HTML codes that *"summarize"* individual web pages for the search engines. With a few exceptions, these codes mostly run behind the scenes, meaning that you don't see them but the search engines do when they crawl your site.

If you want to see what tags are currently in place on your site, open up a page from your site in your browser window and click *"View"* and then *"View Source"* or *"Page Source"* from the navigation menu in your browser. You'll see something open up that looks like this:

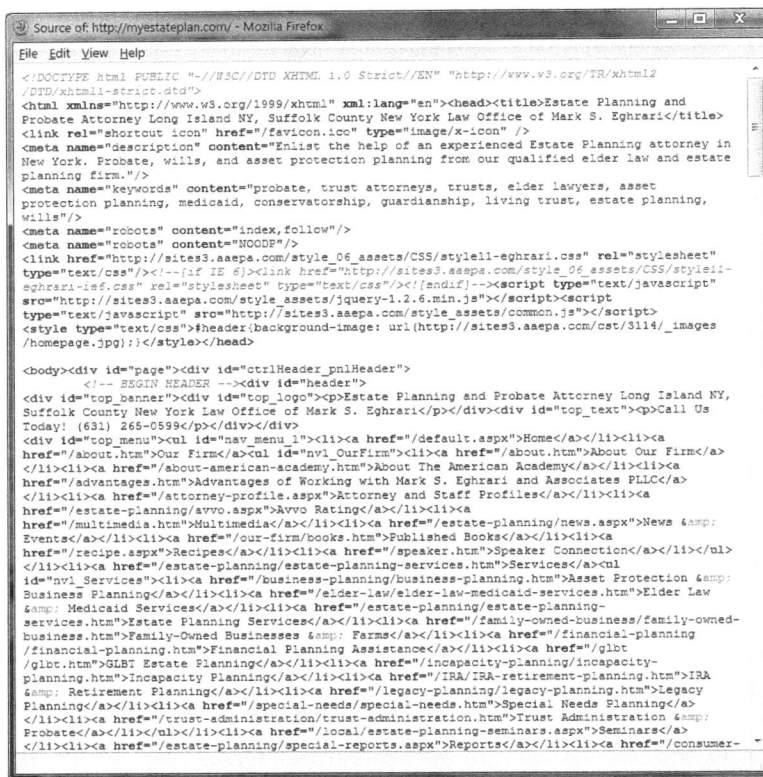

Now, this might look like a bunch of gobbledy-gook at first, but the search engines know exactly what this means. Using the right tags can help make your site more search-engine friendly. There are several different codes you can use as part of your metatags, but the most common—and the ones we'll focus on here—are the title tag, the keyword tag and the description tag.

Title Tag—The title tag is a one-liner that sums up a given web page, both for the user and for search engine ranking purposes. The tag for your home page for example, might read, *"Colorado Divorce and Family Law—The Law Offices of Smith & Davis."* If you want to see what kind of title tags your competitors are using, just look at the top left portion of the browser window when visiting their site. This is where the *"title"* of the webpage appears.

Title tags can also be seen in the search results. If you had a web page on your site titled *"Why You Shouldn't Do Your Own Divorce"* and included this as your title tag, that's what users would see whenever this page made the listings. If you don't use a title tag, the word *"Untitled"* shows up instead.

Title tags should be about ten to fifteen words, give or take; Yahoo! allows for 120 characters while Google allows for only sixty-six. Anything more will be cropped.

Titles are enclosed between the tags <Title> and </Title>, as shown here in one of our Members' source code sample:

Description Tag—The description tag also more than one purpose, helping the search engines rank the page and providing the user with a synopsis in the search results as well.

Using our earlier example, a page titled *"Why You Shouldn't Do Your Own Divorce"* might have a description tag that reads:

> *Handling your own divorce might seem like a good idea at first,*
> *but there are several things you should consider before you do.*

Description tags can be longer than the title tag—twenty-five words is optimal.

Keyword Tag—The keyword tag includes all the keywords that you believe are relevant to the content on a particular web page. Keywords are separated by a comma and no punctuation is necessary.

These tags are used strictly by the search engines for ranking purposes and you should keep your keyword list to no more than ten to twenty keywords per page if you can.

	Title Tag Max Display Length	Meta Description Max Display Length	URL Max Display Length
Google	66-69 Characters	156-158 Characters	75-92 Characters
bing	67-72 Characters	176-184 Characters	91-94 Characters
YAHOO!	64-65 Characters	164-190 Characters	65-80 Characters

[1]

Using the Right Syntax

In order for the search engines to pick up your metatags, they should be inserted into the header portion of your webpage, in between the <Head> and </Head> tags and, **except for the title tag**, are written in the following format:

---

1   *SEOMOZ.org*

```
<Meta Name= "Name" Content: "the meta tag content goes
here.">
```

So, going back to our divorce attorney in Colorado with a web page explaining why do-it-yourself divorce is a bad idea, we might see the following metatags:

```
<Head>
<Title>Why You Shouldn't Do Your Own Divorce—Smith & Davis
Law Offices</Title>
<Meta Name= "Description" Content: "Handling your own
divorce might seem like a good idea at first, but there are
several things to consider before you do. Learn more from
licensed Colorado divorce attorneys, Smith & Davis.">
<Meta Name= "Keywords" Content: "why you shouldn't do your
own divorce, handling your own divorce, divorce laws,
Colorado divorce, divorce attorney, Colorado family law,
Colorado child support">
</Head>
```

As you can see, each tag includes some targeted keywords and provides both the user and the search engines a solid summary for this particular page. Now, let's look at how you use them.

To edit (or create) your tags, you'll need an HTML editor. FrontPage works fine, as does Dreamweaver and many of the other popular software programs. You can also edit HTML in Notepad if you know what you're doing.

Rules for Maximizing Metatags

Rule #1—Metatags are good. But, they're not everything. Remember, search engines have changed the way they measure the value of a website so while metatags are still a good idea, they're not the end-all, beat-all in SEO that they used to be. Think of metatags as your SEO starting point—not the sum of all your efforts.

Rule #2—Keep It Simple. Because metatags used to be held in such high regard, you'll still find many novice webmasters who believe that the more metatags you have, the better ranked your site will be. This is not true. In fact, too many tags can actually hurt your rankings. Less is more where metatags are concerned.

Rule #3—Repeat, Repeat, Stop. It's fine to repeat your keywords a couple of times within your metatags, especially if you using variations of the search phrase however, repeating a keyword over and over and over does not boost your rankings. Instead, many search engines will see it as spam and penalize you for the infraction.

Rule #4—Relevance is Key. Yes, you might want to rank for a variety of keywords. But, if the page in question is about incorporating a small business, then adding keywords that target intellectual property wouldn't be appropriate and ultimately, hurt your rankings. Think about it. If you're searching for specific information, would you want to land on a page that offered no solution to your problem?

Rule #5—Never Copy What Your Competitor Is Doing. You can certainly look! This is considered to be SEO research. You may get some inspiration as a reward for your efforts. But simply copying your competitors' tags doesn't guarantee that you'll attain your competitors' rankings. Besides, you don't want to rank where your competitor does. You want to surpass him.

Rule #6—Different Pages Get Different Tags. Yes, all the pages within your website have a common theme. But, if you want to maximize the benefits of metatags, each page gets its own set of tags. This ensures that each article you write is viewed individually by the search engines and that every page of your site is optimized to the fullest extent.

Anchor Text—Anchor text refers to the word or words you use to link to something else on the Internet. For example, if want to offer your readers a free educational report that explains the different types of business entities, you'd want to add a link to that report on your home page that looks something like this:

```
<a href="www.yoursite.com/freereport.pdf">Click here!</a>
```

In this example, the words *"click here"* represent your anchor text and as it stands, this example would work just fine. But since you're a master marketer, you're going to go one step better and optimize that link for the search engines. So instead of using generic words like "click here, your anchor text should be something more relevant to your site and/or your report. For instance:

"Click here for your free report on incorporating your business."

"To learn more about the right legal structure for your business, get your free report today!"

Once again, these links are much longer, but they provide a clear understanding of what the link points to. And because it's a link, the search engines give this block of text extra weight. In fact, studies have shown that a web page can reach the top of Google's rankings without even having the term in question mentioned on the actual page. It was anchor text from outside websites linking in that gave the page its ranking.

So, knowing how powerful anchor text can be, you certainly don't want all your SEO efforts to net you a #1 spot for the search term *"click here."* Instead, you want to make sure that your links include highly optimized key phrases that support your SEO strategy.

Alternate Image Text—When you insert an image into a web page, you have the option of giving it a title and assigning alternate text. Years ago, this was mostly to appease web users with more primitive browsers that couldn't display the graphics. The alternate text would serve as a placeholder for these browsers, so that the user could see what kind of photo was intended for that particular spot.

Needless to say, not many webmasters bothered to label their images.

Today, however, the alternate text attribute can be a great addition to your SEO efforts. To label your images, your code should look something like this:

```
<img src="image.jpg" alt="Legal Entities Report" />
```

Now, this coding works fine if the image doesn't link anywhere. But, what if you want to include a hyperlink?

Images with hyperlinks should definitely have alt attribute but they should have a title attribute as well. The title attribute is picked up when a user *"hovers"* over the image with a mouse. The browser will display the title as a tool tip.

To take advantage of the title attribute your coding should look like this:

```
<a href="your link here"><img src="image.gif"
alt="Legal Entities Report" title="Get Your Free Report!">
```

Using targeted keywords for your image titles and descriptions, you can boost your page's optimization and thus, your website's measured relevance to your targeted keywords.

Step #3—Building Backlinks

Having created stellar content and fully optimized your site, you're ready for the third piece of the SEO puzzle: backlinks.

Backlinks are basically links from other websites that point to your site, meaning that this portion of your SEO campaign is not totally within your control. Quite often, you're going to have to rely on others to do the linking for you and that can be a hit-and-miss proposition. They are like votes from other websites that confirm your site has good information. Yes, much like a popularity contest.

The weight that search engines give backlinks comes from the notion that, if the content is worthwhile, other web users will link to it—something that's true to some extent but not always a guarantee. We are inundated with thousands of pieces of information every day, So, while we might come across an interesting article or inspirational blog post, we won't always automatically log onto our own website to post the link.

What's more, not all backlinks are equal. You'll remember we discussed page rank at the beginning of this chapter. It's this ranking that gives the backlinks their power.

A page can have a rank of 0 to 10—with 0 being the lowest and 10 being the highest—according to Google's page ranking system. The page rank is calculated by multiplying a percentage of the rank of the pages that link to you by a mathematically-obtained dampening factor. The percentage of the outside page's rank that is used will be determined by dividing the linking page's rank by the number of links on that page.

This equation tells us a couple of different things:

The first is that backlinks are worth more if they come from pages with high page ranks. The second thing this equation tells us, however, is that backlinks from sites with excessive outgoing links are worth less than sites that are more conservative with their links, even if the lesser-linking site has a lower page rank.

Knowing this, we can start to develop a backlinking strategy that will boost our page rank and move us up in the search results.

Who's Linking To Your Site?

Since you're an SEO-savvy webmaster, you'll want a way to monitor your backlinks and the SEO juice they're giving you. One easy way to do this is to use Yahoo! Site Explorer (siteexplorer.search.yahoo.com).

This free service allows you to see how many pages Yahoo! has indexed for your site and also see how many inbound links you have coming in.

Using the American Academy's site as an example, you can see that we have 966 incoming links to the homepage (aaepa.com) as of the date this was written.

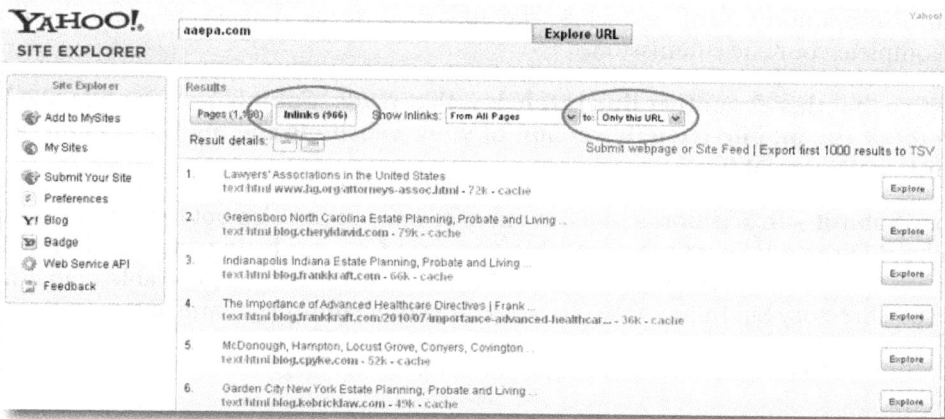

If we change the settings to reveal incoming links to the entire site, that number jumps to 17,711. As you can see, we've been very busy optimizing our web presence.

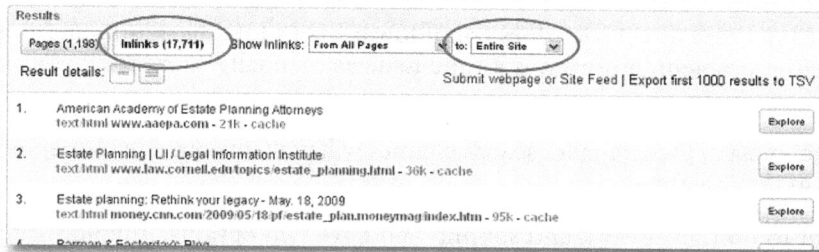

Now, many of these links are coming from our member attorneys' websites. But, you will also see that we've got links coming in from Cornell University, CNN Money and yes, even FindLaw. These links, along with the other seventeen-thousand plus, are what helps the Academy continue to show up on page one of the search engines and dominate the results.

Submitting to Search Engines
For a search engine to start ranking your site, it first has to know that your site exists. After your site is discovered, the search engines will then *"crawl"* your site at regular intervals to update the index of available pages.

But you don't have to wait on the search engines to find you in order to get listed. Just the opposite! Submitting your website to the various directories is an important step in the SEO process. Although submission alone doesn't guarantee instant rankings and it may take a while for the search engines to crawl your site, it is still a good strategy to help your pages get indexed as quickly as possible.

Google—To submit your site to Google, go to **google.com**/addurl/, type in your domain name and the set of scrambled letters shown in the box and then click *"submit."*

Yahoo!—In order to submit your website to Yahoo!, you must first have a Yahoo account and ID. This is free, so first go to **edit.yahoo.com**/registration and create your free account. Your Yahoo! ID is the username you choose when you create this account. Once you've registered for your Yahoo! account, go to **search.yahoo.com/info/submit.html** and click the link that reads *"Submit Your Site for Free."* Follow the instructions to complete your site submission.

Tip: **Did you notice that there's another link just below that reads "Submit Your Mobile Site For Free"? This allows you to submit the mobile-friendly version of your website to the mobile version of Yahoo! Search and, yes, we think that's a great idea!**

Yahoo! Directory Submit—In addition to the free submission, you can also opt to get your listing expedited by choosing the Yahoo! Directory Submit service. This service is $299 per listing and an additional $299 per year to maintain the listing. You're site will be reviewed within seven days and you'll be able to maintain your listing online. To start the Directory Submit process, go to **ecom.yahoo.com**/dir/submit/intro.

Bing—To add your site to Bing, just go to **bing.com/docs/submit.aspx** in your web browser and follow the simple instructions.

What About Other Search Engines?
While the Big Three Directories—Google, Yahoo! and Bing—are certainly the most popular, they're not the only fish in the virtual sea. Quite the contrary, there are literally thousands of search engines out there, although none with the same power or presence as the Big Three.

That said, getting listed in these directories does count toward your total backlinks and is generally considered a good SEO strategy. The problem, however, is a time issue as manually submitting your sites to thousands of search engines just isn't realistic.

Also, keep in mind that most of these smaller search engines will pick up your site sooner or later, especially after your website is listed in the Big Three.

But, if you really want to do the legwork and submit, you have two options: The first is to enlist the help of a submission service to reduce the amount of legwork you and your staff have to do. The second option is to focus only on the top ten, as listed here:

- **Google**
- **Yahoo!**
- **Bing**
- **Ask.com** (formerly **AskJeeves.com**)
- **AOL.com**
- **AltaVista.com**
- **AllTheWeb.com**
- **Lycos.com**
- **HotBot.com**
- **Excite.com**

Internet Directories
In addition to search engines, you should also submit your website to various Internet directories.

The most popular—and also the most beneficial—is **DMOZ.org.** However getting listed in this directory can take a long, long time because each entry is reviewed by a real person before it's added to the site. Still, it has a page rank of eight, so if you do manage to get listed, it's definitely a good backlink to have. You should also add your website to Google Places, Yahoo! Local and City Search, a process we'll cover in the next section.

Blog Directories

If you have a blog—and again, you *should*—then you'll want to submit your blog to the top blog-oriented directories. Like search engines, there are tons of directories to choose from. Rather than bog you down with site submissions, let's just focus on the top four:

BlogCatalog—blogcatalog.com

Registration is free and once you've set up your account, getting listed happens quite quickly. In addition, BlogCatalog allows you to edit your profile and interact with other bloggers, making it a great tool for social networking as well.

Bloggernity—bloggernity.com

Also free, Bloggernity has a quick registration form (user ID, password and valid email address) and currently has thousands upon thousands of blogs listed.

Technorati—technorati.com

This is a must for any blogger. Technorati is perhaps one of the biggest blog directories on the web and its community of bloggers is quite active, meaning traffic generated from this site is worth quite a bit. Registration is free and you can submit multiple blogs to one profile.

Bloggapedia—bloggapedia.com

This community-based site also provides a nifty little users' toolbox that can help you improve your blog, start a podcast and generally enhance your marketing efforts.

Legal Directories

As an attorney, you also have the benefit of getting your blog/website listed in the various legal directories across the Internet.

FindLaw.com is probably one of the biggest such directories. To list your firm and your website, go to **flcas.findlaw.com**/rpu to register and create your profile. Other general directories include **LegalDirectories. com, Martindale-Hubbel (martindale.com)** and **TheLegalDirectory.org.**

In addition to the general directories, there are also several practice-specific directories that you should consider. For example, estate planning attorneys can pay to have their listing featured at Search-Attorneys.com, our own search directory created exclusively for estate planning, probate and elder law firms.

Are these practice-specific directories worth your time? Well, consider that **Search-Attorneys.com** consistently ranks in the top three listings for the search term *"estate planning attorney"* on Google (as does the Academy, by the way—Did we mention the Academy?) as shown in the image below:

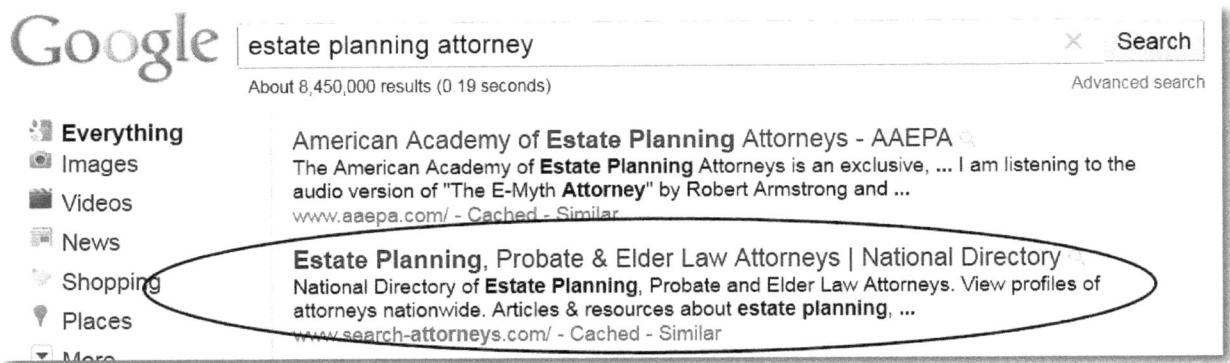

When deciding which directories to use, a quick search can tell you just how much power that directory really has.

For a more comprehensive list of directories, see the Appendices section of this book.

Getting Links from Other Webmasters

If webmasters are inundated with information as we suggested earlier, what chances do we have to convince them to link to our website?

Actually, a very good one in many cases. Just like you, other webmasters know the value of incoming links so they're often willing to place a link on their site in exchange for you doing the same on yours.

The thing you want to avoid is exchanging links with sites known as **link farms**. These sites often promote **link exchange programs** where you both agree to post links to the other's website. The problem, however, is that link exchange programs typically don't offer quality content. So you're building backlinks with a website that doesn't offer any real value. In addition, search engines frown on this practice because its designed to circumvent the algorithms used to measure value. If you get caught, you could be banned from the search engines altogether.

Besides, we already know that the quality of your backlinks outweighs quantity. So, it makes sense to be a little picky about with whom you're exchanging links.

What else should you avoid when developing your backlinking strategy?

- **Exchanging links from a site not indexed by the search engines.** It is possible to keep a page (or set of pages) from being indexed by the search engines. To do this, you create a file known as **robot.txt** which tells the search engines which pages to avoid crawling. If you exchange links with a webmaster and your link is on a page not crawled by the search engines, then the link doesn't do you any good. To check the value of your backlink in this regard, you can run the page through a **search engine simulator** to determine if the page can be crawled.

- **Exchanging links with** *"bad neighbors."* Your backlinking motto should always be: *"take care who you exchange links with."* The reason is that while the search engines won't penalize you for those incoming links to your site (You have no control over who links to you!), they will penalize for linking out to spammy sites, sites with no value, banned sites or other bad neighbors. Check the value of a site before you agree to link to it to ensure that you won't be doing more harm than good.

- **Using an image link when a text link option is available.** While images are certainly more eye-catching, text links rule in terms of search engine optimization. If you have the choice between a text link and an image link, choose the text link every time.

- **Getting a text link with useless anchor text.** We mentioned the power of anchor text earlier but it's worth repeating again here. Several years ago, George W. Bush's presidential profile showed up in the #1 spot on Google for the search term *"miserable failure."* Interestingly, however, neither of these words was mentioned even once on the page in question. The reason his profile ranked for such horrible search terms is that so many others had linked to his profile using those exact words. This phenomenon is referred to as a **Google Bomb**. Since that time, Google assures the public that they've fixed the algorithms to ensure it won't happen again. But that doesn't mean that the anchor text used in your incoming backlinks won't influence how your site is ranked. To be on the safe side, make sure all your backlinks use optimized anchor text.

Building Your Own Backlinks
Now, in addition to requesting links from other webmasters, you can also create your own. The most common way to do this is to participate on other websites, such as forums and related blogs. You can typically include a link to your site when you post a comment at other sites, thus creating a backlink and building your brand at the same time.

Other ways to build your backlinks include article marketing, videos and press releases, all of which we'll discuss in detail in part three of this book.

NoFollow vs. DoFollow Links—What's the Difference?
Every link on the Internet has the ability to serve as a backlink for the website to which it is pointing. But, that doesn't mean that every link fulfills that service. Some webmasters regularly employ a linking tactic known as *"NoFollow."* While there is some debate over its effectiveness, the result is that you could possibly lose some backlinks in the process.

Here's how it (theoretically) works:

If you want to link to something but don't want to lend any page rank in the process, you simply tell the search engines not to follow your link. To do this, you add the following code to your hyperlink: rel=*"nofollow."* Links without this tag are referred to as *"dofollow."*

Now, there are several reasons that you might want to make a link nofollow. Many bloggers sell advertising on their site in the form of links for example, and the search engines frown upon providing backlinks to paid ads. To continue earning your advertising revenue while staying in the good graces of the search engines, you could make all advertising links nofollow.

The problem is that some webmasters and bloggers use the nofollow tag more frequently. Some make all blog comments nofollow, meaning that if you comment on another's blog and leave a link, it may not actually be providing the page rank juice you think you're getting.

Likewise, if you're trading links with other webmasters, you want to be sure that the links they're providing are without that nofollow tag.

Now, whether or not all search engines honor the nofollow tag is still up for debate. But, given that's it's not hard to verify, it makes sense to include this little step as part of your online marketing process.

There are two relatively easy ways to do this. The first is to use the *"View"* function in your web browser. Visit the page that contains the link you want to check, click *"View"* and then *"Page Source."* A new pop-up screen will open that shows you all the coding running on the page in question. Scroll down to find your link and see if the nofollow tag has been inserted.

```
<a href="www.SmithDavisLaw.com/" rel="nofollow">
The Law Offices of Smith & Davis</a>
```

Now, for those of you who don't like to read coding, you can download a little Firefox plugin that will highlight any nofollow links on the page. Go to **www.quirk.biz**/download-searchstatus and download the plugin called searchstatus. This means you'll have to use Firefox as your browser to see the nofollow links but, given that it's free as well, that shouldn't be too hard to do.

Final Thoughts on the SEO Process

While these three components make up the majority of your SEO efforts, you should always remember that the technology is changing rapidly and so will the strategies that utilize it.

Posting comments in forums might be a good idea today for example, but tomorrow these links may have little to no effect at all. To ensure that your search engine strategies continue to pay off, you'll have to stay informed about the latest technologies and Internet marketing strategies.

One final thought: If you're using Internet Explorer as your browser of choice, you might want to rethink that decision. Most serious Internet professionals use Firefox, an open-source browser from Mozilla, as their primary browser because it offers a wide range of tools and plugins to maximize your SEO research.

With Firefox, you can install plugins that allow you to instantly see a site's pagerank, for example, or block popup ads or manage your Gmail account. Firefox is Netscape-based so it's also more secure than IE and it has a powerful cache system that allows it to load pages faster. To download and install Firefox for free, go to **mozilla.com**/en-us/firefox/.

9.2: *Local vs. National Search*

While the idea that you could be ranked in the #1 spot for *"tax attorney"* might sound fabulous, your target audience isn't the World Wide Web. It's just the World Wide Web in San Francisco, Dallas or wherever your neck of the woods might be.

This is known as **Local Search**, a concept we explored briefly when we talked about drilling down your keywords. At this writing, local search is quickly becoming even more important for small businesses than ranking on a national level. In fact, local search is strictly for brick-and-mortar businesses with a local audience. And that's exactly what your law firm is.

So, let's talk about some other ways to maximize your local results.

Building Local Backlinks

In addition to including your location in your optimization efforts, you can also use your backlinking strategy to improve your local rankings.

Getting authority websites in your area to link to you for example, is a good way to connect your business to your geographic area. Your local Chamber of Commerce would be a good link to have as would links from local charities, businesses, and educational websites. The more local links you have, the more the search engines will see you as a local authority in your practice area.

You should also ensure your website is listed in some influential local directories, such as: Google Places, Yahoo! Local and Bing Local. Let's go through these one by one:

Google Places

You've probably already seen Google Places in action and don't realize it. When you perform a search on Google, it reads the location of your ISP provider and returns results that are targeted to your location. So a search for *"bankruptcy attorney"* in Atlanta will get a different set of results than someone performing the same search in Rhode Island. More importantly, Google returns a block of *"mapped"* locations and, these locations ALWAYS show up on page 1.

In fact, local results used to be limited to five and Google positioned them about midway on the first page. Now, that limit has been extended to seven and Google displays local results *above* most of the organic search results!

These mapped locations come from listings in Google Places, which has been aggregated from a variety of local information sites like CitySearch, Yelp and yellowpages.com. In fact, Google created about fifty million of these Places sites as a starting place for local businesses. Google now wants these local brick-and-mortar enterprises to claim the site and customize them with verified information, photos, videos, coupons and client reviews.

Currently only about 3% of business owners have claimed their free spot in Google Places, even though that free spot means, potentially, showing up on page one of the most popular search engine on the web.

Of course, we're smarter than that, aren't we?

To claim your page, go to **google.com**/lbc and log in with your Google ID.

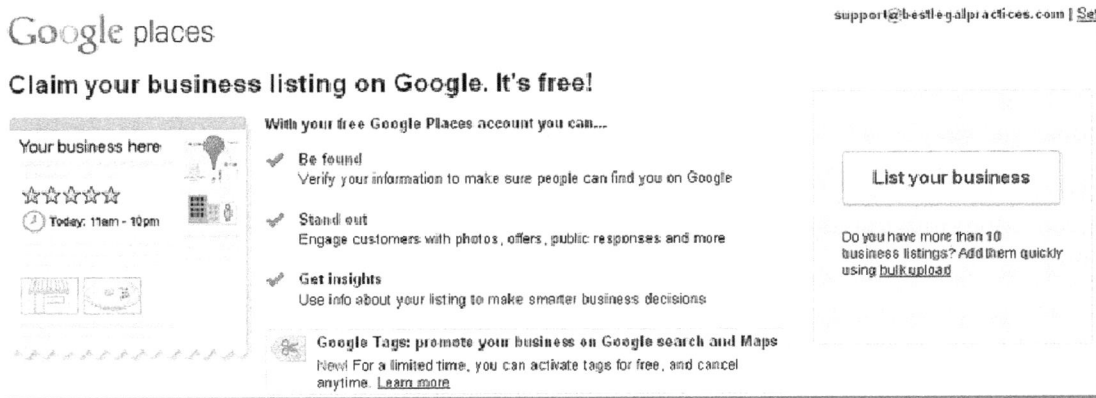

If Google Places already has your firm listed in their maps, they'll let you know and give you the opportunity to edit your listing and verify ownership of the business. If you're not listed, you'll be taken to a page that looks like this:

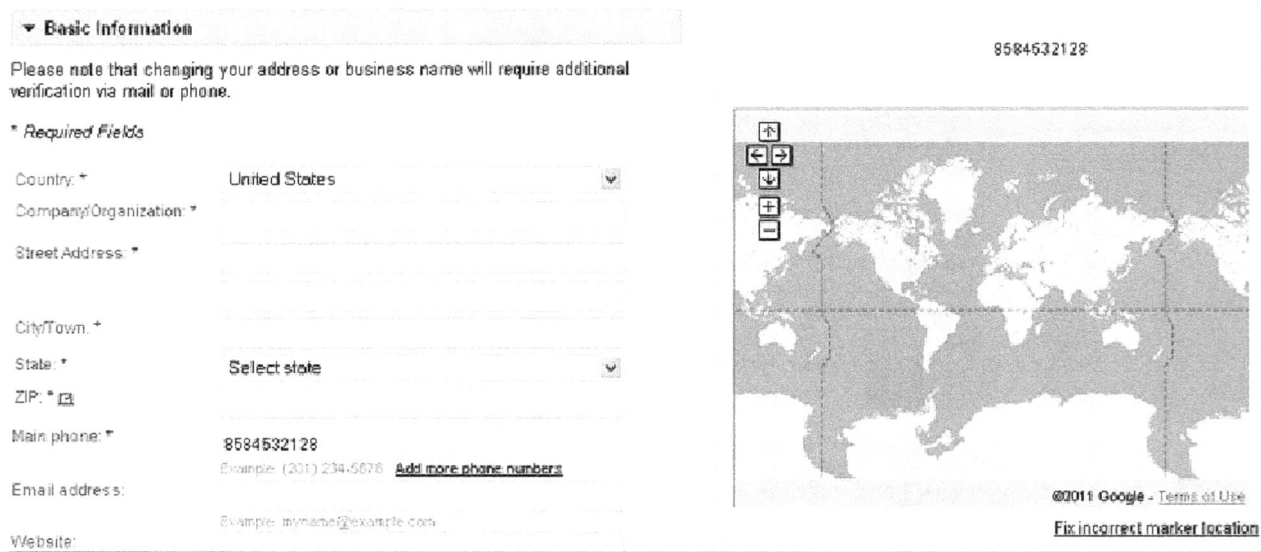

As you complete the information, Google will mark your spot on the map and you can correct the marking if it's inaccurate.

In addition to basic information such as your firm's name and address, you also have the ability to list your website, a description of your firm and a category you'd like to be listed under.

Scroll farther down and you'll find a place to list your hours, payment methods, details about your firm you'd like your prospects to know and upload videos and images relating to your practice.

As with other pieces of your online marketing strategy, Google Places is social in nature. So be sure to include your videos and photos and complete your profile as much as possible.

When you're done, click the *"Submit"* button at the bottom of the page. If this is a new listing, you'll need to confirm that you're the person entitled to claim this listing. To do this, you'll be presented with two options: a Google representative can contact you at your listed phone number or they can send your verification by snail mail, which will take about two to three weeks.

If you opt for the phone call, it will happen immediately, so be ready. Once you receive your confirmation PIN, you'll want to log back into Google Places and enter the PIN in your Google Places Dashboard:

Now, once you've activated your listing, there are a couple of other things you can do with Google Places.

Citations

Citations are essentially *"mentions"* at other websites, similar to backlinks and when your firm is mentioned or reviewed, the Google algorithm eventually picks it up and shows it in your Google Places page. And just like SEO and backlinks, the more/stronger citations you have, the higher up you'll move in the Google Places listings.

So, how do you get citations?

The first thing you need to do is get your business listed in as many directories and sites as possible, so start with those that allow you to create your own listing. Keep in mind that local directories are considered to be more valuable in terms of citation strength, so start with those and work your way out. Look at your Chamber of Commerce, local business directories, local newspaper websites and even other professionals in your area. Forums and message boards work well too. To find these sources, do some local searches that include your practice area—such as *"Boston divorce."*

Unfortunately, you can't submit these citations to Google as you get them. You have to wait for Google to pick the citations up naturally and as you might have guessed, this can take some time. To get the most benefit from citations, continue to seek out new places where you can list your firm. The more citations you have, the better your ranking will be.

One final caveat: Google Places matches information in the citations to the information you provided about your firm, so be sure to use the same address, phone number and email address on all your listing submissions.

Tags

Tags are a fee-for-service option that Google provides to its Places users. This tag alerts visitors to your Places page that you have a special offer or feature they should be aware of. This could be a discount coupon for example, or a new service you want to promote. The tag is displayed as a yellow marker next to your listing in Google Places and users can click on the tag to see your special offer:

To create a new tag, login to your PIN-verified Google Places account and click *"Create Tag"* in the status column of your dashboard. Then click *"Add a Tag"* (top of the page) and follow the directions for adding your tag. When you're done, click *"Add Tag"* to submit.

You can choose from different tag types. Some require more information than others: photos, videos, coupons and posts. Now, posts are short, micro-updates you can add to your Places listing. You can use up to 160 characters (by comparison, Twitter is 140) and, at this writing, you can have only one post at a time.

The beauty of tags is that you can use them to highlight your marketing material. You could add a tag to promote your coupon for a free consultation, for example. Or, you could add a post-tag, sending users to the landing page for your free educational report. Did you create a new video? Then you should have a tag. Did you upload new pictures from your recent open-house? That's worthy of a tag too.

Tags cost a flat fee of $25 per month and can be cancelled at any time. At this writing, Google is offering a free thirty-day trial for new tag customers. So, once you've verified your Places account, you can sign up and try the tags feature out for free.

Coupons

Coupons allow you to offer your clients discounts straight from the web. You can create a coupon by logging into your Google Places account and clicking the *"Coupons"* tab on your dashboard. Click *"Add New Coupon"* and fill in the information. Click the *"Submit"* button. Viola! You now have a coupon!

Your prospects can see this coupon when they visit your Google Places page and incidentally, you can also create a QR code to point to that coupon and help promote it. For attorneys, the coupon could be an offer to receive an initial consultation, free of charge.

QR Codes

A QR (*"Quick Response"*) code is basically a 2D bar code that can be scanned by a wide variety of smart phones. The QR code then takes the viewer directly the your firm's website. Google first released these QR codes as part of a *"kudos"* program for businesses that were ranking high in Google Places. Dubbed the Google Favorite Places Program, these businesses received a window decal that alerted visitors the company was a Google favorite:

Visitors could then scan this QR code with the camera in their smart phone and immediately see the company's Places page, download coupons, read reviews and even leave their own review. (If you scan this code for example, it will take you to Google.com.)

And while there's no strict timeline for when another wave of these decals might go out, you can still create your own QR code from your Google Places dashboard and include it on your business card, your email signature, your website, newsletters and anything else you can think of that would help promote your business. You could even include your QR code in your invoices with a request for your clients to let you know how you're doing by leaving a review on your Google Places page.

To create your own QR code, log into your Google Places dashboard and click the *"Print QTR Poster."* This will generate a *"We're on the Map"* poster that includes your QR code.

Google Hotpot
Is your Google Places listing really all that important? Well, playing off Google Places is Hotpot, Google's answer to other local rivals such as Yelp and Foursquare.

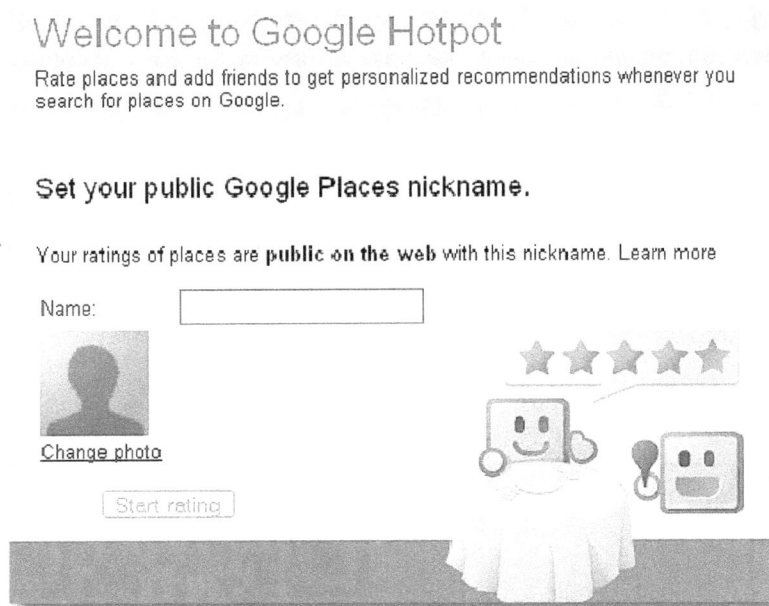

Hotpot allows users to instantly access Places from their smart phones, search for products and services in their local area and yes, leave reviews. That means that a new client could have a review posted about your firm before they ever get out of your parking lot. It also means that the more you can make your Places profile stand out, the better you'll do against the firm down the street.

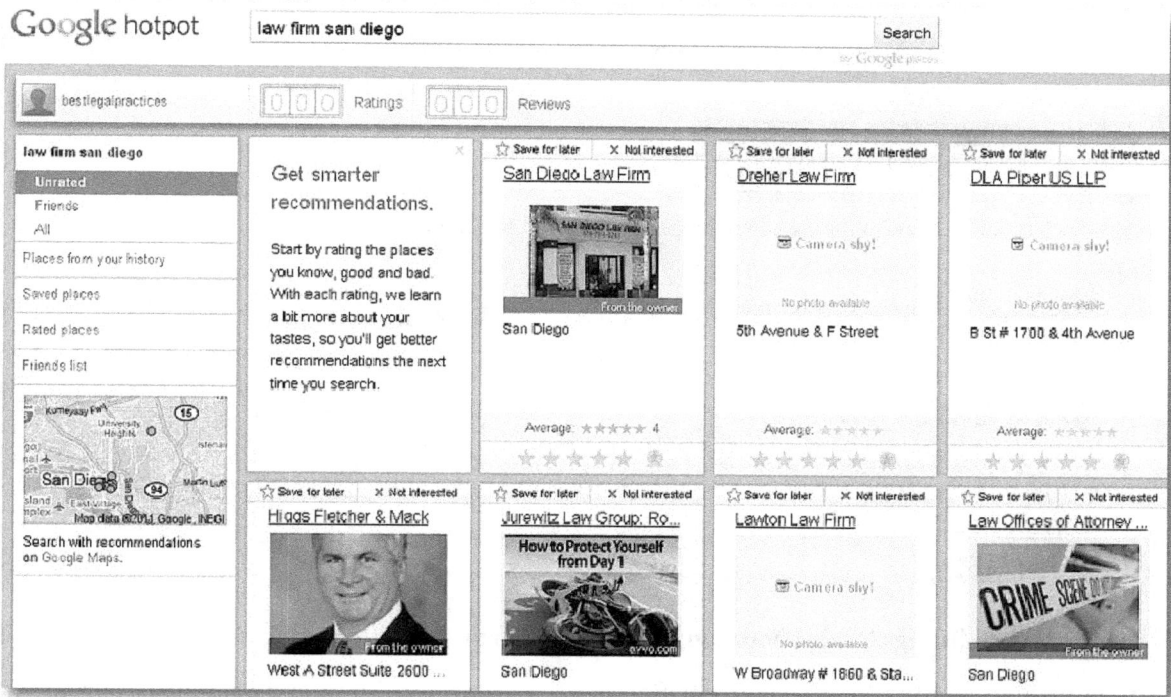

Now, you don't have to do anything with Hotpot. Your listing will come from your profile on Google Places. Just know that it's there and with Google already having apps in place for the Android and iPhone, it's only a matter of time before Hotpot becomes the marketing *"hotspot"* for local businesses.

Yahoo! Local

While Yahoo! doesn't assume that you want local results right off the bat like Google does, it does make it very easy to get those local results if you want them. Users can choose to surf the web and see generic results or they can choose *"Local"* and enter the world of Yahoo! Local.

Yahoo! Local works the same way as Google Places, providing mapped listings from Yahoo! Local's database.

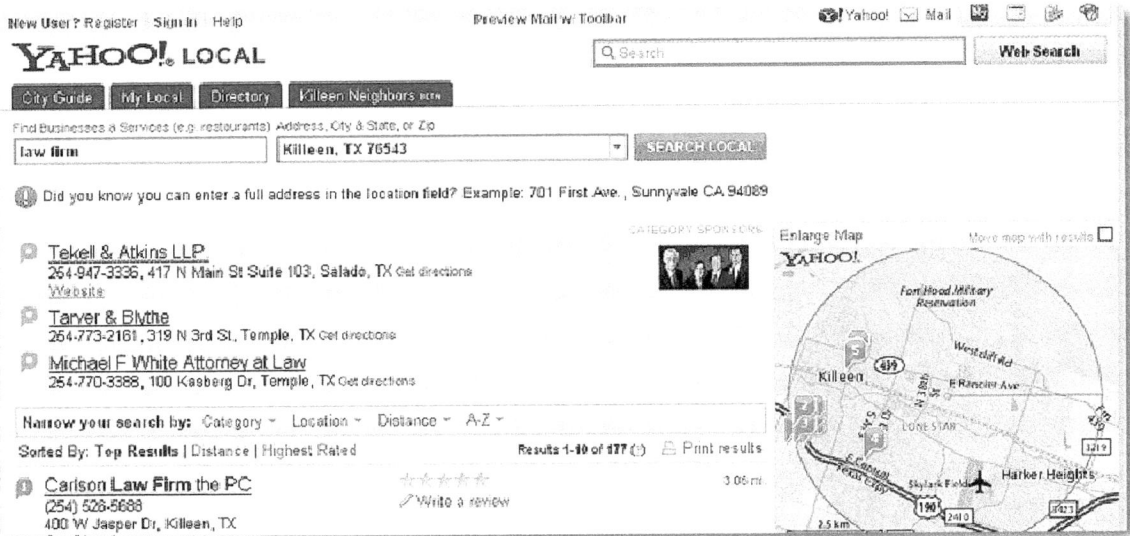

To add your firm, go to **listings.local.yahoo.com** and choose from the basic, free listing or the paid, enhanced listing.

Basic listings include your firm's contact information as well as your store hours and a summary of your services. An enhanced listing will cost you $9.95 per month, but allows you to add your logo, tagline, ten photos and a more detailed business description.

Incidentally, these same listings feed the main Yahoo! page when users add a local modifier to their search, so someone searching Yahoo! for a law firm in Killeen, TX will see the same results that they'd get if they started their search in Yahoo! Local:

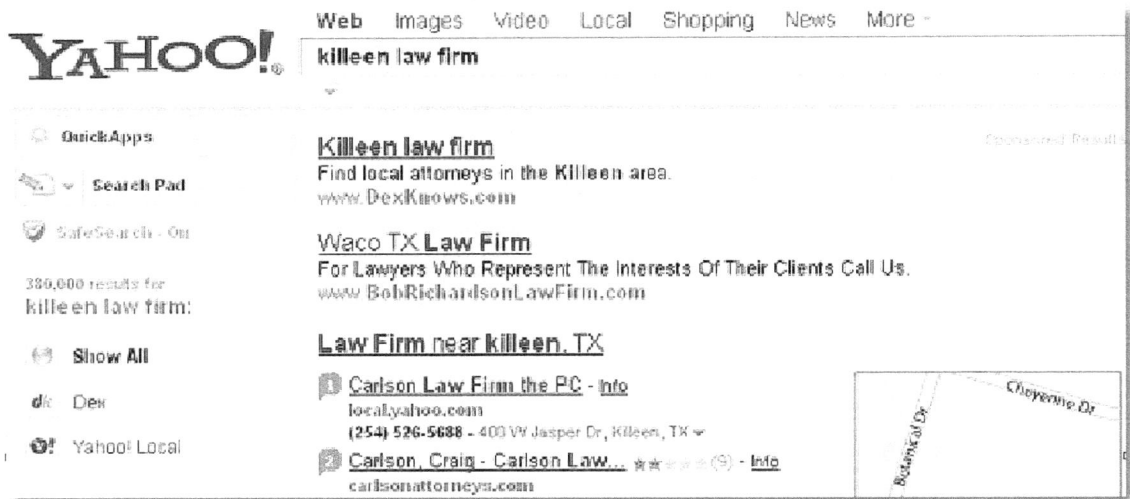

Bing Local

Bing is like a Yahoo! and Google Hybrid. Enter a search term on Bing's home page and you'll get local, mapped results at the top and then broader, generic results below. Bing doesn't automatically assume you want only local results as Google does, but it does incorporate local data in the first page of results.

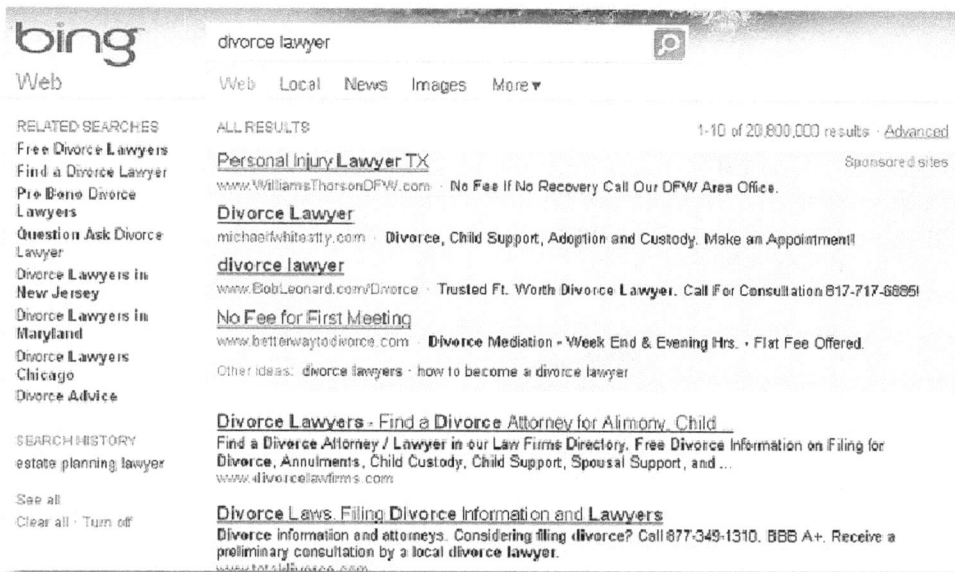

If the results you see here aren't what you were looking for, you have the option of clicking the *"local"* tab at the top to use Bing's local search. Sponsored links come from **yellowpages.com** and are shown at the top and bottom of the page.

To get your firm listed at Bing Local, go to the Bing Local Listing Center at **ssl.bing.com/listings/ListingCenter. aspx.** Click *"Add a New Listing"* and then complete the information requested. Bing will check to see if your listing already exists and allow you to choose the categories where you'd like to see your listing included. You'll need a Windows Live ID to complete the submission—if you don't have one, Bing will give you the opportunity to create one for free.

Focus on Relationships

Now, you might have noticed that the local optimization process involves quite a few of those offline, face-to-face tactics we know and love and there's a reason for that. In local search, relationships are everything. Local search is all about giving web users a reliable way to find the products and services they need, when they need them and what better way than to allow the community where the business resides to determine which companies are best at providing those services and products?

This is inbound marketing at its best: connecting with the client by becoming easily accessible, something law firms haven't always done in the past.

But we're looking forward now, aren't we?

And the future requires a new kind of thinking. This type of local, inbound marketing gives new life to the old referral system we've relied on for so many years. Clients can quickly post thoughts about their experience with your firm and given the power of local search, you should encourage them to do that every chance you get.

Of course, this technology also makes it easier to post a bad review about your firm, so if you haven't yet considered how you'll make your practice more client-friendly, now would be the time. Is a bad review really that detrimental to your firm? Just look at what a few clients had to say about this divorce attorney and decide for yourselves. (Incidentally, these reviews are now forever part of his Google Places listing.)

Reviews by Google users   Been here? Rate and review

A Google User - May 23, 2010
★★★★★ **Awful Services** This guy was more into his ego and who he "knew" in the city than helping me. This is definitely NOT the guy to see!

2 out of 2 people found this review helpful. Was this review helpful? Yes - No - Flag as inappropriate

Oct 29, 2009
★★★★★ **Not Recommended** I do not recommend this attorney: his lack of knowledge concerning my case, his unethical billing practices not to mention the fact that he continued my case multiple times eventhough we wanted to settled via a divorce agreement out of court. He took a very simple case and made it complicated for his financial gain.

Was this review helpful? Yes - No - Flag as inappropriate

More reviews by Google users (2) »

Getting Your Clients' Reviews

Now that you know why you need reviews, the next question is: How do you get them? How do you convince your clients to start writing reviews about your firm?

The best way is to simply ask. Many clients may not realize that you've claimed your spot on sites like Google Places and Yelp. So, letting your clients know that you've joined the online revolution is a good way to entice them to start writing reviews.

Personal requests for reviews can be made via email, on the phone or, of course, in person when you're meeting with your client. But you can also *hint* to your clients by promoting the concept of providing feedback. Create a special page on your website that explains how much you value your clients' feedback and where they can go to leave it. Using the list we've provided in the appendices section (see Local Search Directories), you could provide clients and site visitors with a linkable list of places to leave reviews.

Also mention your love of feedback in your client newsletters, your client contracts and, yes, in the signature of your emails.

Use the good reviews to promote your site and give kudos to reviewing clients at the same time. Post these reviews on your website, on your fan page and in that newsletter you're sending out as well. If the review comes from another business, give them the opportunity to enjoy a little free PR by including a link to their website with their review of your firm. Let other clients see how appreciative you are of the incoming reviews and they'll be much more likely to join in the fun.

What about bad reviews?

It stands to reason that you may have a few bad reviews at some point and time. It's impossible to please everyone and, sadly, people are more likely to leave a review when they're mad at you than when they love your work.

But before you start pretending those harsh words don't exist, consider the opportunity it offers. This may be a chance for you to repair a client relationship that didn't work out the first time around. Contact the client if possible (offline, of course) and see if there's a way to undo the damage. If you do manage to resolve the issue, ask the client if he/she would mind amending the review.

You won't be able to solve every issue, of course, and the more positive reviews you have, the less impact those few negative words will have. Just remember that every review, good, bad or otherwise, contains valuable information. Learn from it and your firm will prosper.

3 Principal Ways to Boost Local Search Results

1. Optimize your listing with current information, photos, videos and coupons.

2. Keep adding new citations from other websites. It should be an ongoing project.

3. Acquire and post as many client reviews as possible on your legal services and client experiences. (Be sure to check with your state's Bar ethics rules for permitted reviews.)

So, besides the three big directories, where else can you post reviews to boost your local presence? We've created a list and added it to the Appendices section to get you started. But, remember, the web is constantly changing. So what's hot today may be gone tomorrow. The directories we've put together are good resources as of this writing. You'll need to continue to do your own research to stay on top of new tools and trends.

Chapter 10:
Setting Up Shop

Now that you know the plan, let's start laying out the groundwork. You need to decide which pieces you have, which pieces you need, and then get the basics set up *before* you start marketing.

So, let's take a look at what you currently have in place.

10.1: Taking Stock

The first step in launching a new marketing campaign is to assess what you've already got in place. This helps you decide what you can work with and what still needs to be done prior to your launch.

Now, since we've already established that your website will be your base of operations when it comes to online marketing, let's begin there.

Is Your Website Worthy?
The first question we need to ask is: *Do you have a website?* If the answer is no, then this should be the first thing on your list. You'll need to purchase your domain name and your hosting to get started. We'll deal with design and logistics in just a bit.

If you do have a website already in place, then the next question is: *Is it functional?* By this, we mean does it already have a blog installed? Does it load quickly and properly represent your firm's brand?

Does it have content? Is it well written? Is your contact information prominently displayed throughout your site and can your site visitors easily navigate from page to page?

Your job at this point is simply to look at your website with a critical eye. Don't settle for *"good enough."* If you aren't sure, then have someone else be the judge. In fact, have several people go over your website with a fine tooth comb and ask them to be as honest with their feedback as they can possibly be.

Mailing List
Every firm—whether you realize or not—has a mailing list. At the very least, it will consist of your existing clients but hopefully, you'll also have contact information for potential clients.

Whatever you've got, gather it up and see what information, if any, is missing. Do you have email addresses for everyone on the list? If not, can you get them? When you've collected all your contacts, have the list converted into a spreadsheet, using a separate cell for each piece of information, i.e., first name, last name, address 1, address 2, and so on or have them entered into your firm's database.

Marketing Materials
We're going to create a whole new set of marketing materials for online use. But there's no reason to reinvent the wheel if some of this material is already done. Pull out your brochures, reports, pamphlets—whatever you've got—and see how they fit with the newly-updated branding concept you've created. You may find that you already have some of the content you need for your marketing campaign.

While we're scrutinizing, now would be a good time to take a look at your business cards and letterhead. They may need some updating to add in your new website address, tagline, logo and the like.

10.2: *Choosing Your Site Design*

We've said it before and we'll say it again: your website is your virtual home base. It's the place where all your prospective clients will end up at some point. It should be, without a doubt, the most interesting and informative piece of your online presence.

The good news is that there are plenty of tools and applications to help you make your website pop. Readers can download data, watch videos, access audio files, send you emails, comment on your content, and read your articles within seconds of you posting them.

This *"everything and the kitchen sink"* technology means that your readers are not going to be impressed by a semi-functional static website that your file clerk created when you just had to get something up. Likewise, they won't be patient with you if the pages load slowly or if the files they're hoping to access have been moved or no longer work.

Now, as the layout goes, you have quite a bit of flexibility here. Feel free to create something that adequately captures your firm's mission, values, and image. Just be sure to follow these five simple rules:

5 Things Every Website Should Have

There is no doubt that every website is unique. In fact, that's exactly what you want yours to be. But, before you put your site out there for the world to see, you should know that there are some basic rules that every website should follow, regardless of the industry, specialty, or topic.

Professional Design
It should probably go without saying but if you don't know what you're doing, then don't. Your website's design should be clean, eye-catching and most certainly professional. These are elements you can get only from an experienced designer. If you happen to have a staff member in your office who's capable of giving you this kind of design quality, then by all means, utilize his/her talents. But if you're cutting corners because you don't want to spend the money on a professional designer, you'll only be hurting yourself. There is nothing more painful than to visit an accomplished attorney's website, only to find that it's disorganized, distracting or, well… tacky. Frames are out, as are big pop-up windows and free hosting. Get someone who knows what he/she is doing from the get-go and save yourself a big headache later. Remember, this is the virtual home of your law practice. So you should be willing to spend what's necessary to create the proper image.

Good Navigation
No matter how big or how complex your website, it must have good navigation. And this should begin with your core pages: home, about, contact, blog, etc. So many attorneys, wanting to give the appearance of having a cutting-edge site, opt for the flash menus with multiple levels of content. This is fine if you have the content to fill all those levels, but it should still be relatively easy to navigate. Basic pages, such as your contact information, should be linked on every page of your site and very easy to find.

Light on Dark
While many logos look great on a darker background, you want your actual content to be contrasted against something very light, preferably white. The reason for this is simple: Reading white text on a dark background

actually strains the eyes and your readers will find themselves wanting to click away, even if they're not sure why. If you just absolutely must have that dark background for your logo, then use the dark color for your header and contrast it with a lighter background for your content.

Interaction

Static websites are out so if yours isn't doing anything, it's time to revamp. This can be as easy as adding in a contact form or you could look into a plugin that pulls your latest Twitter tweets and Facebook status updates into your website. Polls, surveys and reader comments are also all good ways to add some community to your site.

Up to Date

There is nothing worse than stumbling onto a site and discovering it hasn't been updated in months or worse, a couple of years. Blogs make your lack of involvement with your site obvious because they show the date of each entry, but even non-blog sites can look outdated if the information they offer clearly should have been changed by now.

For example, a tax attorney's website that still features tax changes for 2008 obviously needs some updating and has probably been forgotten by its owner. Likewise, a divorce attorney's website that is still citing 2007 divorce statistics is probably due for an overhaul. Not having a website is one thing. Having one that you've neglected for years is much worse.

Choose Your Graphics Carefully

While some strategically placed graphics can do wonders for even the most basic of websites, you have to be carefully about the size of the images and the number you choose to use.

Graphics are measured in kilobytes and the more kilobytes your graphic has, the longer it takes to load. Does that mean you can't use a few high-resolution images on your site?

Absolutely not! Just use them sparingly. A page full of images will take forever to load in a browser window. The result is that your prospective client will go looking elsewhere.

Also remember that the people coming to your website are there for a reason. Chances are, it's not to check out your graphics. Use images that support your overall theme and/or that illustrate a particular point or topic, but balance those images with plenty of solid, useful information.

Since we're on the subject of images, don't use graphic backgrounds. It's cheesy. It's cluttered, and it distracts from your content.

10.3: The Beauty of Blogs

In the beginning there were simply websites: basic, static pages that were designed solely to provide information. Users would look up your website, find your contact information and call (or visit) your establishment to get more details. These were nothing more than fancy, online brochures.

To put things in perspective, a restaurant that included its menu online was considered to be cutting-edge.

Now, however, users not only expect to see the menu but they also intend to order—right from the web. That's because today's average website contains a variety of bells and whistles, including e-commerce capabilities, file-sharing, document storage and interactive platforms, otherwise affectionately known as Web 2.0.

One of the most popular applications in this arsenal is the blog. Short for *"web-log,"* blogs allow publishers to create and manage an on-going database of content that can be sorted in a variety of ways, i.e., via categories, keywords and tags. The blog's content is typically presented in reverse-chronological order so that the user always see the most recent content first, unless of course, they search for something specific.

Blogs can offer several benefits:

- Search engines love content—the more the better—so by continuously adding new content, you're also constantly improving your rankings.

- Users love to comment. One of the features of blogs is that you can allow your readers to comment on a particular post. Build a reputable blog and you'll find that your readers leave comments and even have full-blown discussions about your topic on a regular basis.

- Blogs are user-friendly. When you're searching for information on a particular topic, blogs are a great resource because, chances are, there's more than one article to address the various aspects of the topic in question. And because this content is managed in the blog's platform, it's both easy to access and easy to navigate.

- Blogs are owner-friendly. Not only are they relatively easy to setup and maintain, but blogs also provide a great platform for you to toot your own horn. Did you just hire a new associate or win a big jury award? Your blog would be a great place to discuss your law firm's latest news.

Blogs also seem to have some impressive influence over potential clients. During a recent survey, HubSpot asked its readers how much blogging affected their decision to purchase. The response was significant. 60% agreed that blogs often had an effect on their decision to purchase (or not purchase) a particular item or service.

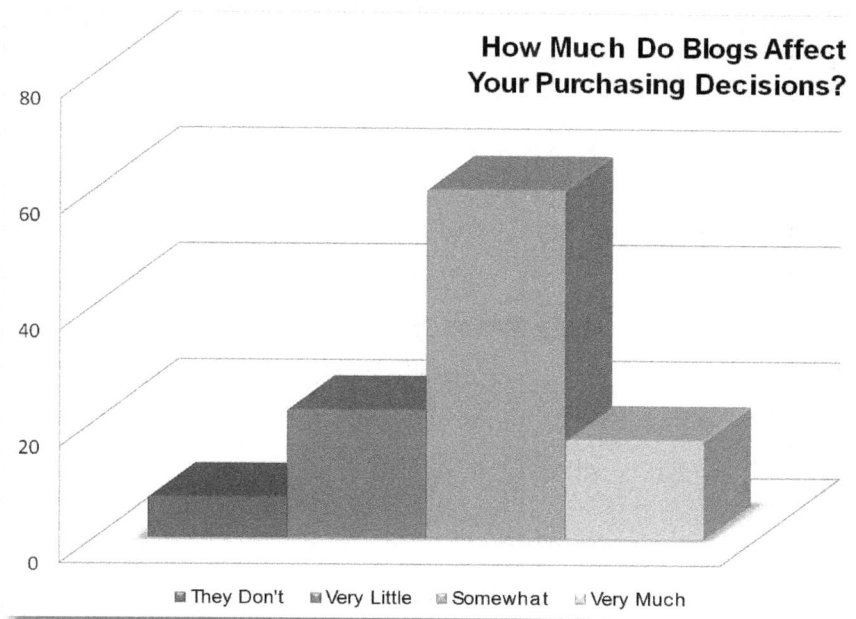

What's more, companies that blog on a regular basis have 434% more indexed pages on the search engines than similar companies that don't blog. Want some more statistics to put blogging into perspective?

When comparing companies that blog to their non-blogging counterparts, companies that blog frequently...

- ...have 97% more inbound links

- ...have 55% more website visitors

- ...are 57% more likely to acquire a client

- ...receive 67% more leads on a monthly basis

Obviously, the moral of this little story is to blog and blog often. So, does that mean you should have a blog instead of a website?

Actually, you need both. Instead of looking at blogs and websites as two separate options, considering using them together. Your website can still host your basic pages, contact forms, attorney profiles and the like while your blog would be a component of your website, dedicated to providing a continuous flow of content, news, and firm updates for your readers.

Fortunately, this kind of setup is not as difficult as you might think. Blogs have come a long way since they first hit the scenes in the late 90s. They can accommodate a variety of layouts and templates.

---

1   *HubSpot—The Science of Blogging, www.hubspot.com/the-science-of-blogging/*

One blogging platform in particular—WordPress—is considered to be one of the most flexible and engaging content management solutions available on the market today. With a good designer/developer, you can create a website that meets all your needs and is built entirely on the WordPress platform.

Plus, WordPress is Open Source. This means that the platform itself is free. Your only expense would be having your designer/developer customize the layout and functions of the site.

Have we convinced you to add a WordPress blog to your website yet? Good, because we're going to spend this section telling you how to get the most from your blog.

How WordPress Works

WordPress is built using a programming language called PHP. This language is very versatile and allows you to create complete websites that can do amazing things.

Let's look at some of the features that WordPress has to offer:

Schedule Ahead
To get the benefit of having a blog, you need to make regular entries, but that's not always easy given the type of schedule that lawyers typically maintain. To ensure that your blog content is published on a regular basis then, WordPress includes a scheduling feature that allows you to post your content to your blog and then *"drip"* it at a later date. This means that the blog post stays hidden until the time and date you designate. So, as far as your readers are concerned, you just made a new post—even if that post was actually created days and even weeks or months in advance.

Plugins
Because WordPress is such a popular platform, users and developers have gone to great lengths to create add-ons that enhance the user's experience. Known as plugins, these modifications can do everything from install an e-commerce platform in your existing blog to customize a category of posts so that they stand out as feature promotions.

Plugins can be used to affect the layout of your site, add new features and enhancements for your readers to enjoy, or even just simplify and/or boost your administrator's interface.

Social Promotion
Using third-party applications, you can integrate your social media accounts into your blog so that your readers can see what you're tweeting about or what you're posting on your Facebook page. You can also include quick links to your various social profiles to encourage your readers to follow you everywhere.

Multiple Authors
WordPress now supports multiple blogs and multiple authors. So you, your partner, and other staff members could blog individually, for example, or you could assign different topics to your associates. You can also limit access to a blog, thus giving your subscribers exclusive content that no one else can get.

Posts vs. Pages

To add something to the blog portion of Wordpress, you simply create a new post. This post is then added to the ongoing database of content and displayed on your *"blog"* page as the most recent addition.

But some content isn't really meant for a blog-type of layout. Your About Us page for example, is considered static content—something that you want to be accessible always with just a click of the mouse, not information that gets pushed farther down as new content is added. Enter the pages feature of Wordpress.

Pages allows you to create new content and give it a static location within your website. Your About Us page, as we just mentioned, plus your attorney profile pages, services page, and other primary content pieces are perfect for the pages feature.

Categories vs. Tags vs. Keywords

When you get into the WordPress interface, you'll see the option to add three different types of classifications to your posts. These are: categories, tags and keywords.

Now you're already familiar with **keywords** and you choose them for a given post just like you would for your website. If your post is about new legislation that affects tax rates, for example, then your keywords would include phrases such as: tax laws and tax legislation as well as more defined key phrases such as: *"what's your new tax bracket"* and *"how the new tax laws will affect you in (year)."*

Categories, on the other hand, are the sub-topics that will determine how your posts are sorted. You can create as many categories as you need. But, keep in mind that good navigation should always come first. So, create your categories with common sense.

Using our example above, the post on new tax laws could be filed under a category called *"Legislation."* It can also be filed under a category called *"Personal Finances."* Or, if the tax laws affect corporations and other business entities, you could file it under a category entitled *"Business Taxes."* The good news is that you don't have to choose. You can file your post under as many categories as you choose and you can even set up a *"General"* category to contain posts that don't fit anywhere else, such as those about events and happenings in your law firm.

Tags are a weird little element in that they're sort of categories and sort of keywords. If your WordPress theme doesn't have a place to enter keywords (and some don't) you can just plug them in as tags and you're done. Life is simple and all is right with the world, once again!

But many themes feature extra SEO plugins, most of which include the keyword feature, leaving us wondering what the heck to use for our tags.

You can still plug your keywords in as tags or you can use tags as another level of sorting your posts. The keywords entered are for SEO purposes only. They won't be used by WordPress as a sorting option. Tags, on the other hand, will. WordPress allows your readers to sift through your posts by selecting categories and tags. So, if you've written that post on new tax legislation, it makes sense to have a tag called legislation or new laws or something similar. If you want to utilize all three (cats, tags, and keywords) you could reserve categories for major groupings (legislation), add tags to break that big group down (tax legislation, business legislation, etc.) and then add drilled-down keyword to attract your target audience (How will the new tax legislation affect your small business?). Get the idea?

Configuring WordPress to Name Your Posts

When you install WordPress for the first time, there are a number of configurations you might want to make before you begin posting. Installing your plugins first is always a good idea and so is modifying the way WordPress names your posts.

Unless the theme you're using has already done it for you, WordPress defaults to a naming sequence that goes something like this: your URL/?p=##, where number is the sequential number of the post. These are called **permalinks**—i.e., permanent URL links to your posts. They're assigned automatically by WordPress as you create a new post.

So, your insightful post on new tax legislation might be named smithdavislaw.com/?p=46. It's not really the best naming strategy for SEO purposes, is it?

Fortunately, you can change the format of your permalinks relatively easily.

Go to the *"Settings"* option in the left-hand navigational menu (You may have to scroll down.) and then choose *"Permalinks."* You'll be presented with several options:

Common options:

○ Default
» http://www.homebizpal.com/?p=123

○ Date and name based
» http://www.homebizpal.com/2006/12/18/sample-post/

○ Numeric
» http://www.homebizpal.com/archives/123

◉ Custom, specify below

Custom structure: `/%category%/%postname%/`

Optional

If you like, you may enter a custom prefix for your category URIs here. For example, /taxonomy/tags would make your category links like http://example.org/taxonomy/tags/uncategorized/. If you leave this blank the default will be used.

Category base: _____

What you want to do is set up a permalink that uses the title of your post to further maximize your site's SEO. To do this, you'll choose the *"Custom"* option and make sure that the text in the *"Custom structure"* box reads as shown in the image above: /%category% / %postname% /.

What this does is tell WordPress to turn your title into a name for the file, so a post titled *"How the New Tax Laws Will Affect Your Small Business"* and filed under the legislation category becomes **www.smithdavislaw.com/legislation/how-the-new-tax-laws-will-affect-your-small-business.php**.

Now, ideally, you want to change this naming structure before you make your first post. If you don't, WordPress won't be able to find your existing posts because they'll already be filed under the old permalink structure.

But never fear! There's a fix. When you create or update your permalink structure, WordPress generates new rewrite rules. It will try to edit the appropriate file so that all your posts are structured properly. This file is called your .htaccess file and is located in your root directory. If WordPress is successful, you're good to go. But if, for some reason, it can't write to this file, it will alert you that you need to correct the file yourself. Again, not a problem, because WordPress also gives you the coding you need to insert into the .htaccess file. Copy the coding. Find the file in your root directory using your CPanel access or an FTP program. Insert the code at the end of the file and save.

Choosing Your Blog's Design

If you already have an existing website with a design you like, you may want to simply have your designer customize the WordPress files so that they fit seamlessly with the rest of your site.

If however, you don't yet have a website or if you're thinking it's time for an overhaul to your design, you might want to consider using WordPress as the base.

With WordPress, you can set up your layout so that your home page looks like a traditional home page should, with links and promotional blocks of text rather than just a scrolling database of articles.

If this sounds like the option for you, there are two ways to go about creating a new WordPress website. The first is to work with one of the thousands of pre-made themes or you can hire a designer to create a customized (and one-of-a-kind) theme just for you.

Now, themes are an easy and typically inexpensive way to go. But one word of warning: many of the free themes (and there are tons) look wonderful but include an encrypted footer file that contains a selection of links chosen by the designer. These are usually links to the designer's homepage as well as various (and often unrelated) affiliate programs that the designer has joined. If you use one of these templates, you can't edit that footer file. It's a tradeoff. You get a nice theme for free and the designer gets the benefit of your website's promoting his or her links.

Creating Your First Post

To create a post in WordPress, log into the administration side, typically found by going to **www.yoursite.com/** wp-admin. This will take you to the WordPress *"dashboard."*

From here, click *"Posts"* in the left hand navigation menu. You'll be taken to a new page that lists all your current posts but now you'll also have a new option under the Post menu link called *"Add New."* Click and you're ready to go.

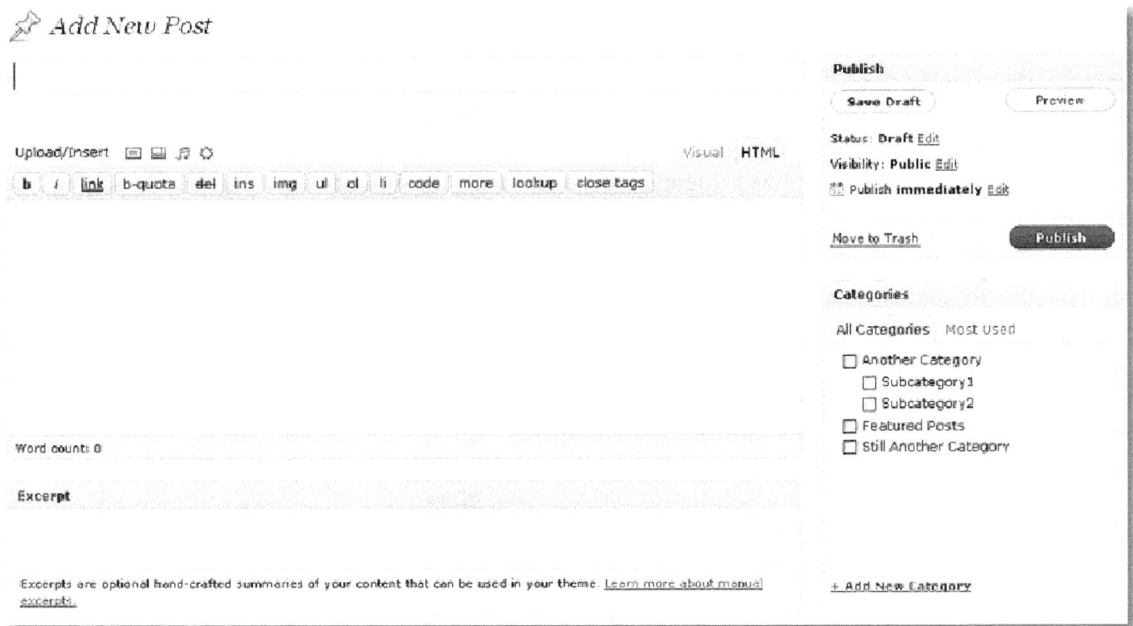

Title

This is the title of your post. If you've configured your permalinks correctly (see previous section) then it will also become the name of your file. Now, this happens very quickly. So don't start typing the title until you're sure you know what you want. Otherwise, you'll need to edit the title manually.

Titles should be written with keywords in mind and should also reflect how a user might search for this particular topic. For example, a post that explains how to collect on past-due child support should be titled something like *"How to Collect On Past-Due Child Support"* or *"What To Do When Your Ex Doesn't Pay Child Support."* These titles are SEO-friendly and they're also enticing to the reader. Someone in this particular situation will think, *"Yes! That's exactly what I'm looking for!"*

And that's exactly what you want.

Post Block

The actual post goes in the large text box that sits just below the title box. You can choose to create and edit your posts using a WYSIWYG interface (similar to what you'd see in your MS Office software products—b = bold, I = italics, etc.) or, if you're feeling adventurous, you can choose to create and edit in straight HTML.

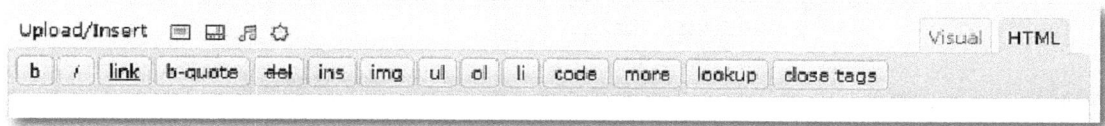

The post can include all HTML components and tags, such as: bold, italic and underline as well as bullet points, tables, and the like.

Publish

To the right of this box is a sidebar that contains a few different publishing options. The first is the actual *"Publish"* box where you can choose to save your post for later, publish, or schedule for publishing at a later date.

Below this box is a list of your current categories. Choose as many as you think would be applicable, or you can create a new category by clicking the *"+Add New Category"* link instead.

Categories

All Categories | Most Used

☐ Another Category
    ☐ Subcategory1
    ☐ Subcategory2
☐ Featured Posts
☐ Still Another Category

+ Add New Category

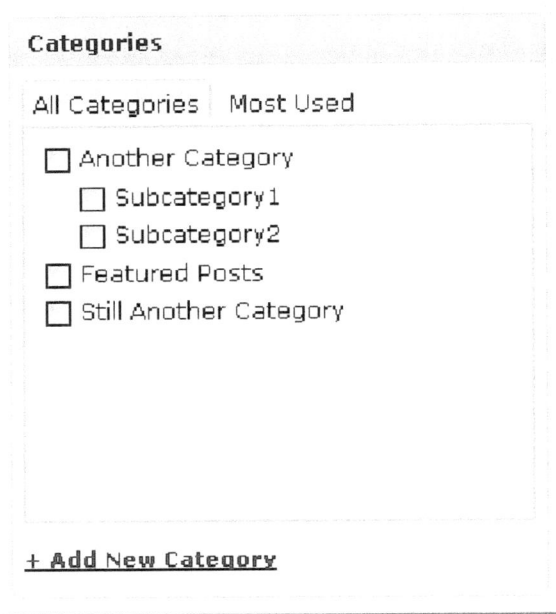

The next box is for your tags. These are entered manually. Type your tags just as you would your keywords, separated by commas, and then click the *"Add"* button.

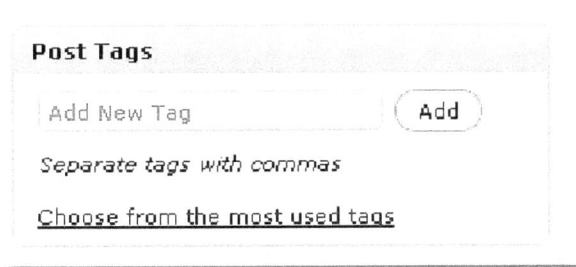

Post Tags

Add New Tag   (Add)

Separate tags with commas

Choose from the most used tags

Once you have your post the way you want it, you can save or publish and you're ready for the next one!

How Often Should I Blog?

In order to get the most benefits from your blog, you need to post on a regular and frequent basis. At the Academy, we recommend posting several times a week to our members and for those who are part of our SEO/SM program, we drip their blog posts on a regular schedule to space them out during the week.

Now this particular schedule isn't written in stone. If you can blog more often, then more power to you. In fact, according to a recent HubSpot survey, the more a company or website posted new content to their blog, the more likely they were to acquire a new client as a result:

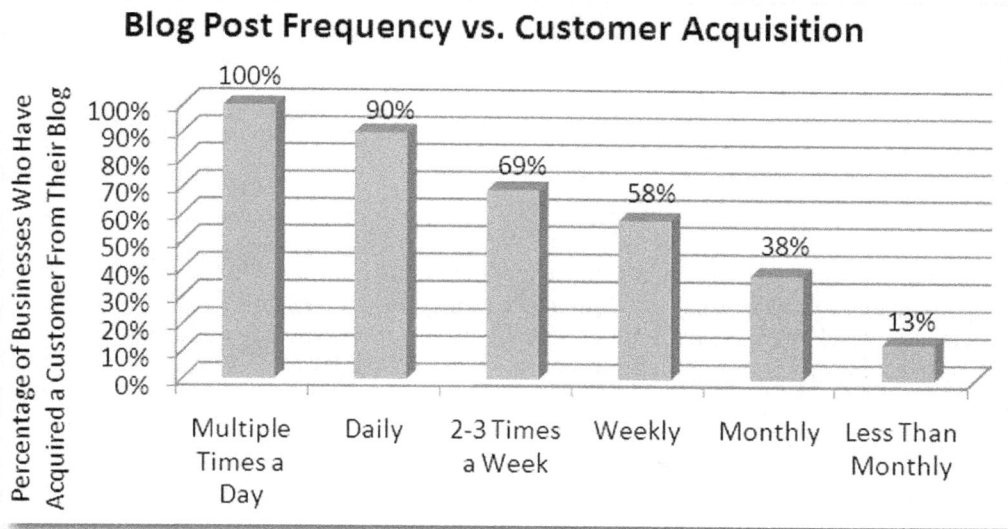

Blog Post Frequency vs. Customer Acquisition

Percentage of Businesses Who Have Acquired a Customer From Their Blog

- Multiple Times a Day: 100%
- Daily: 90%
- 2-3 Times a Week: 69%
- Weekly: 58%
- Monthly: 38%
- Less Than Monthly: 13%

[1]

What's more, blogging seems to help leverage social media for small businesses. As demonstrated in the chart below, small businesses that blog frequently have 102% more Twitter followers than their non-blogging counterparts:

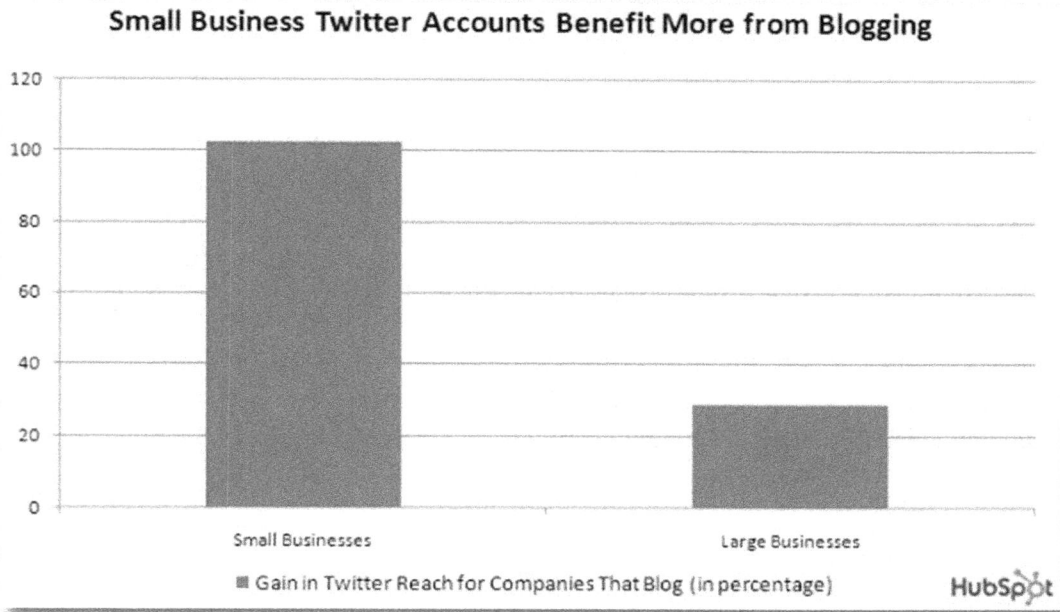

Small Business Twitter Accounts Benefit More from Blogging

- Small Businesses: ~102
- Large Businesses: ~28

■ Gain in Twitter Reach for Companies That Blog (in percentage)

HubSpot

[2]

1   HubSpot—State of Inbound Marketing Report—www.hubspot.com/Portals/53/docs/resellers/reports/state_of_inbound_marketing.pdf
2   HubSpot—Small Businesses that Blog Have 102% More Twitter Followers, blog.hubspot.com/blog/tabid/6307/bid/5459/Small-Businesses-That-Blog-Have-102-More-Twitter-Followers.aspx

Ultimately, your blogging schedule should depend on several things:

Your Readership
If your readers are busy professionals, they may not read as often throughout the week and tend to catch up on weekends. If this is the case, you may see a higher traffic flow on some days than others. You should consider adjusting your schedule to post during your highest traffic times, keeping in mind that you still want to keep a steady stream of new content flowing on a regular basis.

Your Schedule
If you're the only one doing the blogging, a Monday, Wednesday, Friday schedule may not work so well for you. Consider having other people from your firm contribute to the blog as well and you can spread out the schedule among all the participants. Your paralegal for example, could blog on Mondays about relevant news items while your office administrator could post any firm-related news on Tuesdays. Add in a few associates (and yourself) blogging about concerns related to your practice area. Just like that, you've got yourself a very active blog.

Your Results
The great thing about setting a blog schedule is that you can change it whenever you need to. If your stats simply don't support a Monday morning post, then perhaps you could try posting later in the day or skipping Monday altogether. That's the beauty of having all those analytics we mentioned previously. They tell you exactly what your readers are looking for.

What Should I Blog About?

This is probably the question we get asked the most at the Academy. Our answer is always the same: everything!

Blog about the recent additions to your firm, the upcoming seminar you're hosting in your office, the new free report you just added for download and, of course, anything relevant to your practice area.

Your keyword research can help greatly in this last regard by showing you what your target audience is searching for. If an estate planning attorney sees that *"writing your own Wills"* regularly gets a high number of searches, it's a smart idea to write a post on that very topic. Likewise, a divorce attorney might write about collaborative divorce, using mediation, the problem with do-it-yourself forms, and a comparison of joint custody versus sole custody. Using our earlier divorce attorney example, *"getting a divorce"* was a great key phrase to target, so why not create some blog posts around that topic?

Just remember to make your blog personal and casual. Comment on staff achievements, anniversaries and birthdays and include pictures. People want to know, like, and trust their attorneys.

Your objective is to create a website that is so beneficial, so informative and *so very valuable* that anyone looking for information or guidance in your practice area can't help but see your site as a resource. These are the sites that you see at the top of the search engine results. These sites are the ones that web users really just can't do without.

For example, below are two images from Frank & Kraft, one of our member law firms at the Academy. The first image is of their website's homepage and the second is of their blog. Notice how the two are seamlessly integrated? See the blocks of upcoming seminars, news and resources? With today's design and development tools, your website can be and do just about anything. So make it something special. Give your prospects things to do, to read, and to download. Don't just tell them how you *"know your stuff."* *Show them*. And let your website be your showcase.

HOME
OUR FIRM
SERVICES
SEMINARS
REPORTS
RESOURCES
CONTACT US
BLOG
Estate Planning Information Center
FRANK & KRAFT
A PROFESSIONAL CORPORATION
ATTORNEYS AT LAW
Call Us Today!
(317) 684-1100
FOLLOW US ONLINE
RESOURCES
Free Consultation
Request a free, no obligation consultation
Estate Planning Attorneys in Indianapolis, IN
Helping Families Preserve Their Wealth During and After Their Lifetime
Thank you for visiting the Estate Planning Information Center for Indianapolis, Indiana, brought to you by Frank & Kraft, Attorneys at Law. We are estate planning attorneys based in Indianapolis, IN whose mission is to
SEMINARS
Living Trust Seminar
JOHNSON COUNTY, 3/1/2011
02:00 PM - 04:00 PM
Living Trust Seminar
INDIANAPOLIS - NORTHEAST, 3/1/2011
07:00 PM - 09:00 PM

HOME
OUR FIRM
SERVICES
SEMINARS
REPORTS
RESOURCES
CONTACT US
BLOG
Estate Planning Information Center
FRANK & KRAFT
A PROFESSIONAL CORPORATION
ATTORNEYS AT LAW
Call Us Today!
(317) 684-1100
CATEGORIES
Advance Directives
Advanced Directives
Asset Protection
Divorce
Elder Law
Estate Planning
• Pet Planning
When to Update your Will
Author: Marvin J. Frank, Estate Planning Attorney / Category: Estate Planning, Wills and Trusts / Posted: 26 Jan 2011
When creating a Will, it is not a once and done activity. Estate planning, which includes creating a Will, is a process. It is wise to consult with your estate planning attorney every three to five years to update your plan to reflect changes in your life, goals, and assets; changes in the law; and changes in your estate planning attorney's experience. When creating a Will keep in mind that you should contact your estate planning attorney immediately (don't wait for the three to five years to expire) if any of the following are applicable for you:

The Importance of Creating Quality Content

Now, given that content is obviously king, you might be wondering what stops webmasters from throwing up as much content as possible, regardless of quality, to manipulate the search engines and improve their rankings.

The answer is the search engines themselves.

Believe it or not, Google knows if your content is worth the web-space it's written upon and will punish those who try to manipulate the system with less-than-stellar copy. These penalties can range from devaluation of pagerank from incoming links to burying your site in the SERP's, never to be seen again.

And while the search engines aren't perfect in this respect, they are getting better. After noting some low-quality links with high page rank, Google's main guy, Matt Cutts, issued a blog post alerting webmasters that the algorithm would be changing... again.

This time, the change was to target sites that produced large quantities of duplicate content (also known as scrapers) as well as those sites that produced spammy or low-quality content in an effort to manipulate the SERP's.

"As 'pure webspam' has decreased over time, attention has shifted instead to 'content farms' which are sites with shallow or low-quality content. In 2010, we launched two major algorithmic changes focused on low-quality sites. Nonetheless, we hear the feedback from the web loud and clear: people are asking for even stronger action on content farms and sites that consist primarily of spammy or low-quality content. We take pride in Google search and strive to make each and every search perfect. The fact is that we're not perfect, and combined with users' skyrocketing expectations of Google, these imperfections get magnified in perception. However, we can and should do better."

What does that mean for you?

Content is most definitely king. But only high-quality content, please. Anything less, and your online marketing efforts might be all for naught.

10.4: *Creating Marketing Materials that Really Work*

The great thing about online marketing is that you can have several campaigns working at the same time. What's more, each of these campaigns can address different target markets, different issues and concerns, different areas of your practice and even different stages of client development.

A free report on the dangers of dying without a Will for example, is a good way to raise awareness in your target market and pique the interest of prospects that weren't aware of the issue. At the same time, a report that explains the pros and cons of a Living Trust is good material for prospects already researching the topic.

Ditto for whitepapers, blog posts, seminars, and newsletters. All these materials help your prospect understand an issue, solve a problem, or address a need. And sometimes, your materials simply serve to help the prospect realize that the need is there.

To attract your prospects, these materials should be free—free to download, free to attend, free to access. There's no commitment, no obligation to buy. In fact, these marketing materials don't directly *sell* anything. They're designed merely to educate and inform. This is exactly what gives them their appeal.

Now, the interesting thing about this type of marketing material is that, if created correctly, it will actually do the selling for you. Even though this material is designed to be free and, specifically, to avoid the hard-sell, its purpose is to draw your prospects in and convert them into clients. This is often referred to as **results in advance**. You're giving your prospects a solution in advance—before they make any commitments or spend any money. Who wouldn't want that?

To create this type of marketing material, think about your practice area in terms of specific issues. An estate planning attorney might want to focus on the problems with dying intestate, for example, or what to expect when probating a loved one's Will.

By the same token, a tax attorney might want to create materials that focus on ways small businesses can reduce their taxes or explore the differences between a traditional IRA or a Roth.

Remember, these materials are designed to educate and inform. So they should be as specific as possible. In fact, the more specific the better, because then you can create multiple reports and articles to deal with different aspects of a common issue.

Condense your information into lists and bullet points as much as possible. The average web user is conditioned to digesting information in small blocks, something Twitter and Facebook can both attest to. Then, give it a catchy title that piques your prospects' curiosity and compels them to want to learn more. Using our report examples above, we came up with the following potential titles:

- 5 Reasons Your Estate Might Be At Risk

- Secrets to Surviving Probate

- 10 Ways to Reduce Your Tax Bill

- Is A Roth IRA Really Right For You?

For someone who's facing the probate process or wondering how to minimize their taxes, these titles are right on the mark and your prospects are often willing to share their contact information with you in exchange for access to that report. This not only tells you who your prospects are, but also what problem they need you to solve.

Be creative. Think about the questions your clients often ask. Look at articles that relate to your area of practice. Pick your topics and then create powerful, insightful materials that show your readers just how invaluable your guidance can be.

Remember, you don't have to limit yourself to just one campaign. You can use all of these ideas or a combination of the ones that work best for you. A quick look at your practice can tell you which areas to focus on first. What kinds of services do your clients seem to need most? What issues cause them the most concern? What is the number one obstacle holding your prospects back from buying into your firm? Answer these questions and then develop marketing materials that address them head-on.

Part 4:

Dominate!

Chapter 11:
Establishing Yourself as the Expert

Regardless of where your expertise might be, your prospects are all looking for the same thing: information. They want to know the difference between a limited liability company and a general partnership. How can they protect their trademark and what can they do if it's infringed upon? How long does it take to file a divorce? What are the rules for filing bankruptcy now that the laws have changed? How can they stop their partner from taking money from their company?

You can provide this information. In fact, that's exactly what your website, your blog, your free reports, and all your other online marketing materials will do. But there's a difference between just offering some vague observations and really establishing yourself as an expert in your field.

And that's what we're going to explore here.

The online marketplace is looking for a go-to law firm—someone they can trust. They're looking for an expert. While you might be just the person for the job, you have to demonstrate your knowledge consistently to earn the title.

Fortunately, online marketing gives you a variety of ways to do just that. Like the world of offline marketing, you'll need all these lead generation campaigns working together.

11.1: Blogging

We talked about blogging before: what it is, what it does and why you need it. Now, let's talk about how to use it as a tool in your marketing campaign. Blogging should be done regularly and your blog posts should address a variety of issues within your practice area.

A divorce attorney, for example, could write about custody issues, the difference between mediation and collaborative divorce, what to do if your spouse has abandoned your family, and how to collect on outstanding child support payments.

An IP attorney, on the other hand, could blog about the copyright issues facing fashion designers, the importance of registering your trademark, and how to protect your digital media.

Regardless of what you write about, you want to make sure that it's done consistently and that each post is both informative and written for a web audience.

Now, this last one is important because there's a difference between writing for the web and writing for a trade journal. Readers on the Internet respond best to small, digestible chunks of information, written in a casual tone. Use bullet points and sub heading to break up a longer post and don't dance around the issue with a lot of legalese. Get to the point and explain the issue clearly and concisely.

Okay, so we've got your blog going. Now, let's talk about how to promote it.

Blog Feeds

One of the big features of having a blog is that the content you post on it can be syndicated across the web. This means that others can subscribe to this syndication and be notified when you've added new content. Think of it as a variation on the traditional newspaper subscription. Whenever a new paper was produced, a copy was delivered to your door.

The same can happen with your blog.

The most common way to deliver this constantly-changing content is through RSS, an acronym for Real Simple Syndication. RSS converts your content into code called XML, which can then be read by various RSS *"readers."* These readers can be third-party applications or they can be built into software and services that you already use.

Your Yahoo! and Google home pages, for example, offer the ability to add your favorite feeds to your personalized home page. You can also read RSS feeds from your Outlook email software as well. In addition, readers can subscribe to these feeds and receive email alerts within seconds of your posting an update to your blog.

Pretty nifty, right?

You can generate this type of feed with pretty much any blogging platform. But, with WordPress, the feed is already built in. To find out your feed's *"address"* in WordPress, you need to know if you're using custom permalinks or default. (See Chapter 10 for further explanation of permalinks.)

If it's custom, type yoursite.com/feed into the URL address bar on your browser window where *"yoursite.com"* is your blog's domain:

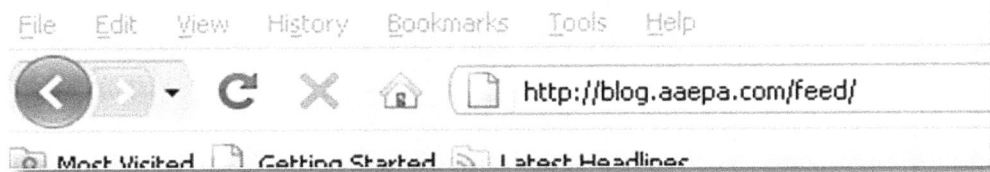

If you're using the default permalinks, your feed URL should be yoursite.com/ ?feed=rss2. With the right URL, you should see something like this:

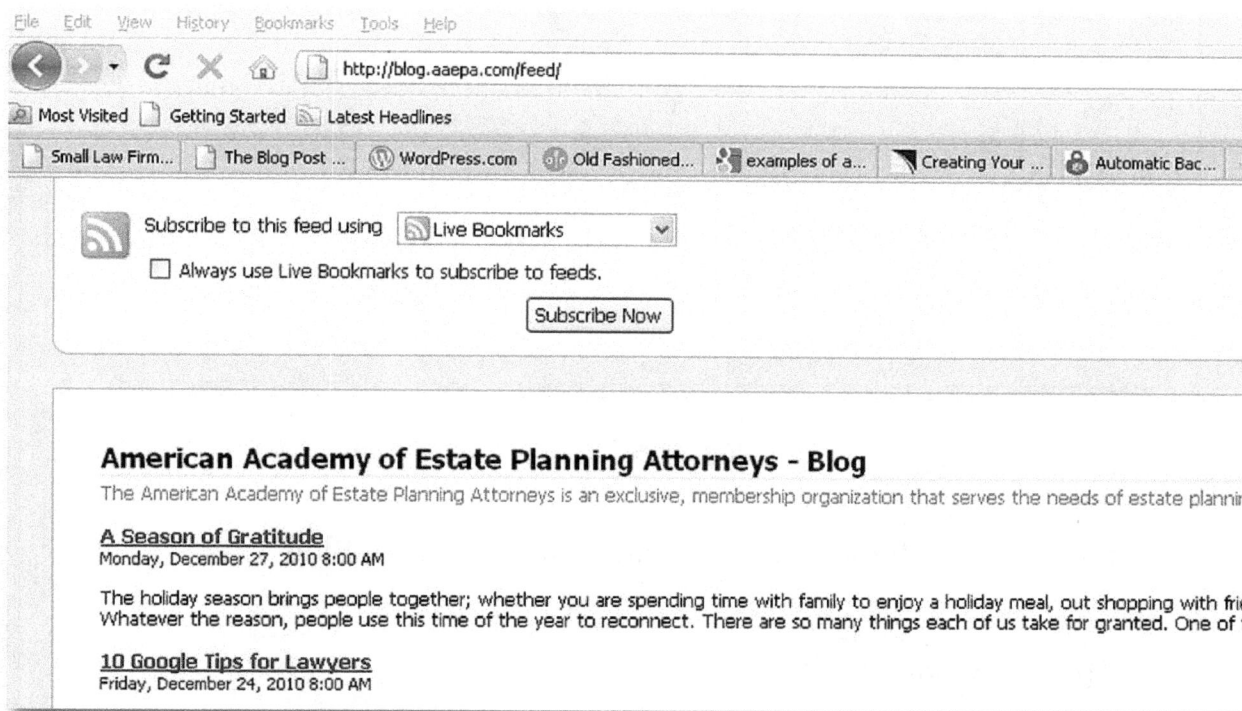

To syndicate this feed, you'll need to add the orange RSS icon——into your website or blog template and link it to your RSS feed URL. As an alternative, you can use a feed syndication service, such as FeedBurner.com, to manage your feeds and monitor your subscribers.

Blog Pings

In Chapter 10, we mentioned the importance of submitting your blog to various blog directories, such as Technorati. In addition to those, you should also enlist the help of a third-party service called Ping-O-Matic.

This service sends out a *"ping"* to a variety of blog engines and directories, alerting them that you've just posted something new on your blog. This ping prompts the blog engines to crawl your blog just as the search engines crawl your website, indexing new posts and feeding it out to the rest of the blogging world.

To use this feature, go to PingOMatic.com, enter your blog's name (Law Office of Smith & Davis—Divorce Attorneys), your blog's home page (**www.SmithDavisLaw.com**/blog or blog.**SmithDavisLaw.com**, for example) and your blog's RSS URL, as we discussed in the previous section.

Check the services you'd like to ping. (Go ahead and choose all the *"common"* services.) Then click *"Send Pings."*

Social Bookmarking

We're going to cover the social media aspect in the next chapter but, in addition to networks such as Facebook and Twitter, you can also promote your content with social bookmarking sites.

Bookmarking allows users to organize and manage the resources they find on the Internet by *"marking"* the page. To do this, the computer notes the URL link and stores it in a folder, typically your *"Favorites"* folder on your web browser.

Social bookmarking follows the same principle except that it adds a social aspect to the process. Not only can you organize and manage your favorites but you can also share them with others on the web and search to see what other people are bookmarking. Not only does this add backlinks to your website, but it's also a good community-building tool.

There are hundreds upon hundreds of bookmarking sites you can use and we've included a long list in the appendices section in this book. As with everything else on the web, there are a few sites worth mentioning separately. Page rank is noted in parenthesis (PR#) and, yes, all of these sites have *"do-follow"* links, so you'll get the backlink juice from Google.

Digg (PR8)

Digg uses a ranking system—measured in *"diggs"*—that allows its users to promote useful or interesting links to the Digg homepage. The more diggs your link receives, the longer it will stay at the top of the page. In addition, links can be broken down by various categories, including: business, science, politics, world news and technology. To monitor your submitted links, you'll want to create your own profile by registering on the site. Go to **Digg.com** and click the *"Join Digg"* button to create your profile.

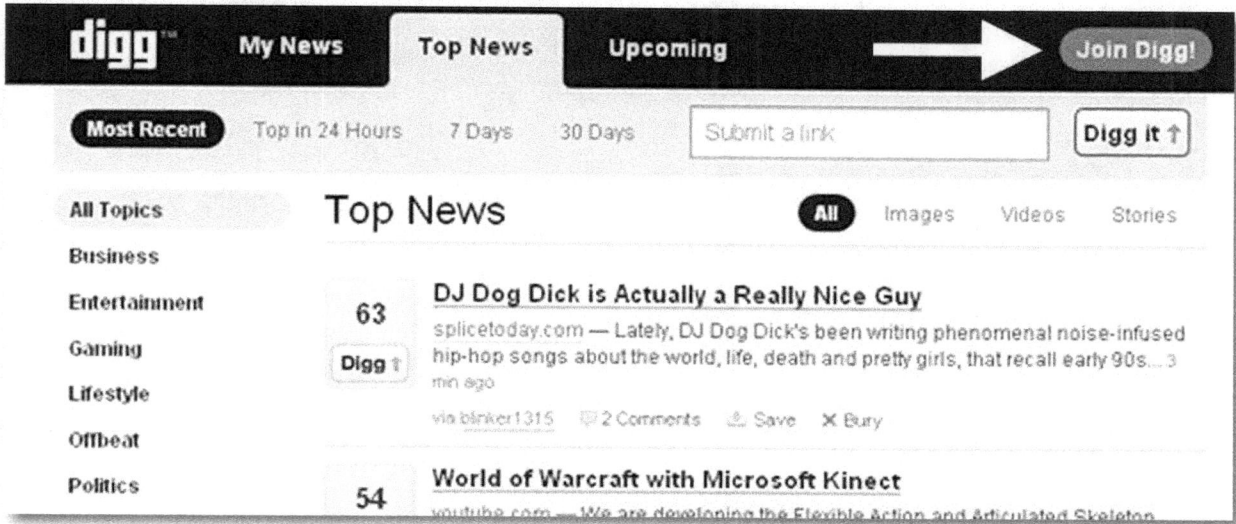

Once you're logged in, you can connect your Digg account to your Facebook, Twitter and Google accounts. (Click *"Connections"* from the left-hand menu.) You can also choose other *"profiles"* to follow by choosing *"My News"* and then *"Find Profiles."* Follow other Digg users by importing your connections from your social media accounts.

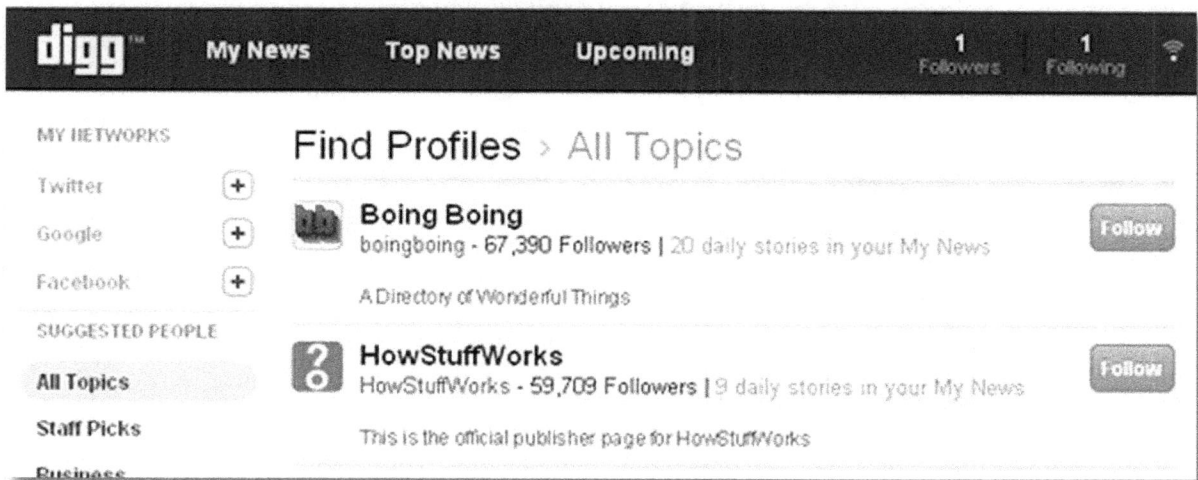

This is an important step because the links you submit show up in your followers' news tab. That means the more followers you have, the more people you'll have seeing your links.

To submit a link in Digg, go to *"My News"* or *"Top News,"* type your link in the text box at the top of the page and then click *"Digg it."*

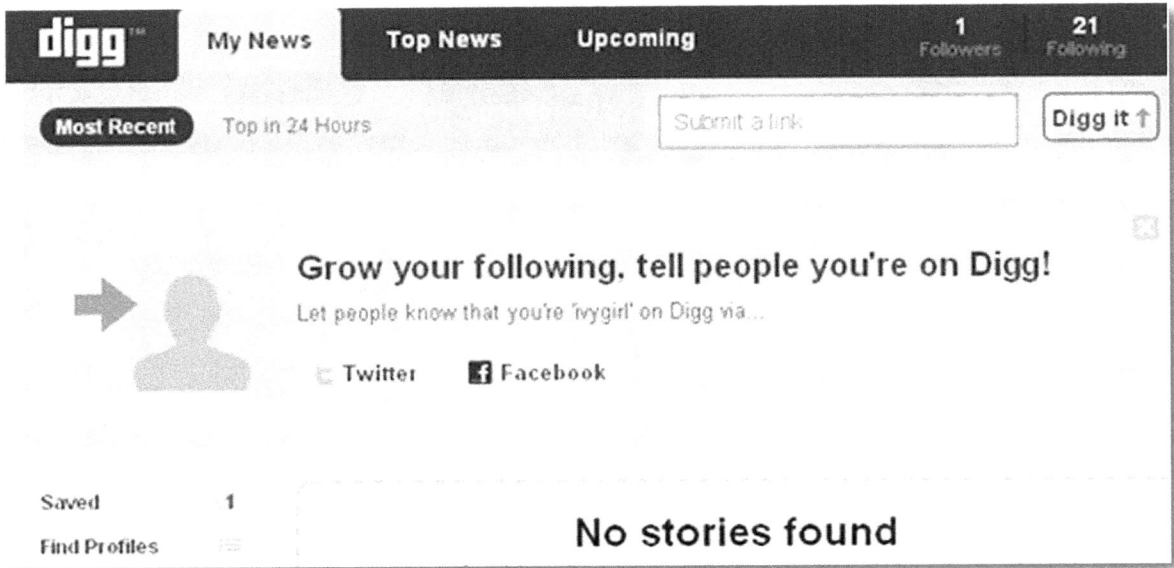

If the story has already been submitted, Digg will give you the opportunity to add your comment and *"digg"* it as well. If it's a new one, meaning Digg doesn't yet have it in its database, you'll be asked to give it a title and description (or edit the default), choose an image, and assign a category (topic) to your story.

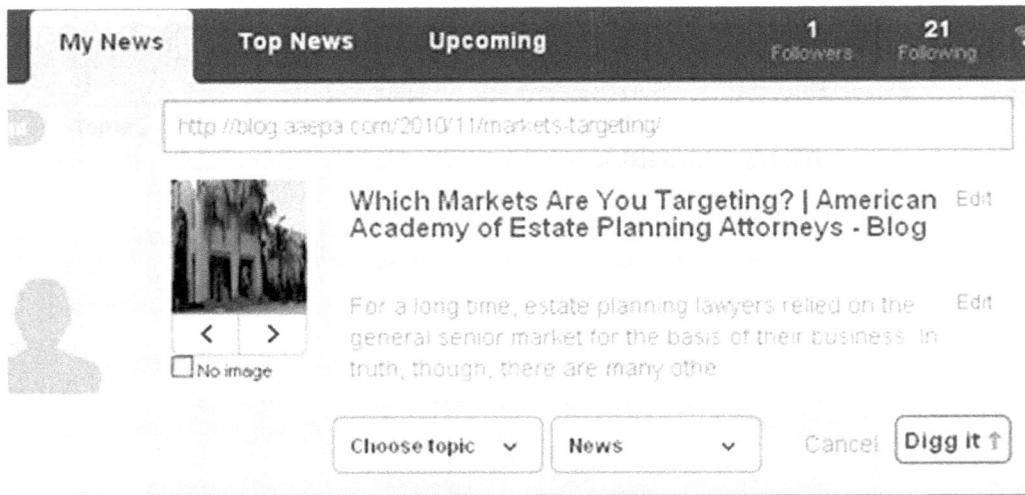

Click the *"Digg It"* button when you're done. You'll then have the opportunity of sharing your new submission on Facebook, emailing it, tweeting it or saving it. If you want to monitor your link's progress in Digg, you'll want to save it so you can view it later.

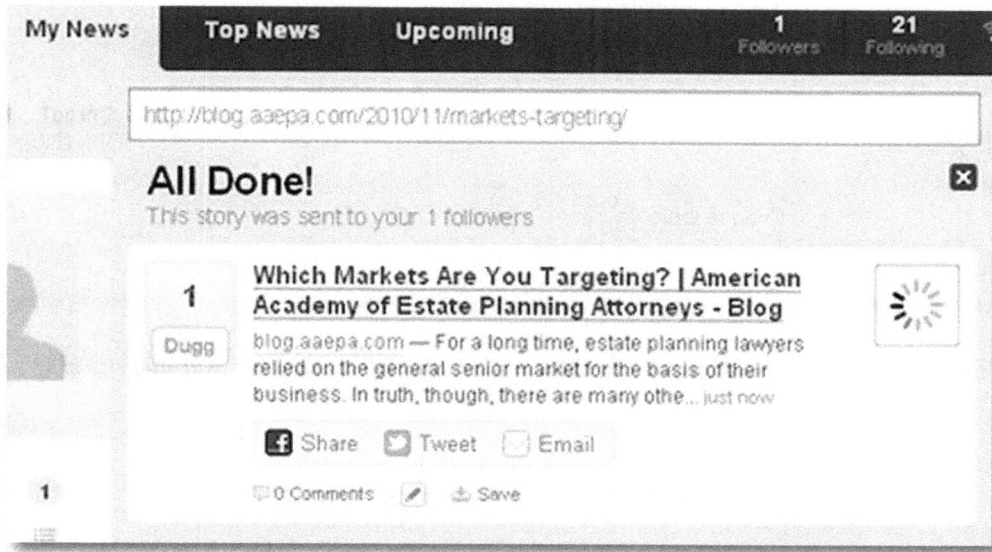

To view a saved story, click the *"Saved"* link in the left navigation menu, just under your profile image from the *"My News"* or *"Top News"* tabs.

Importing Your Blog

Digg now offers its users the ability to import a blog feed, meaning that all your new posts will be automatically submitted to Digg. To do this, go to your profile by clicking the dropdown arrow next to your profile icon at the top of the screen:

On the left-hand side, you'll see a new navigation menu. Click the *"Import Feeds"* link:

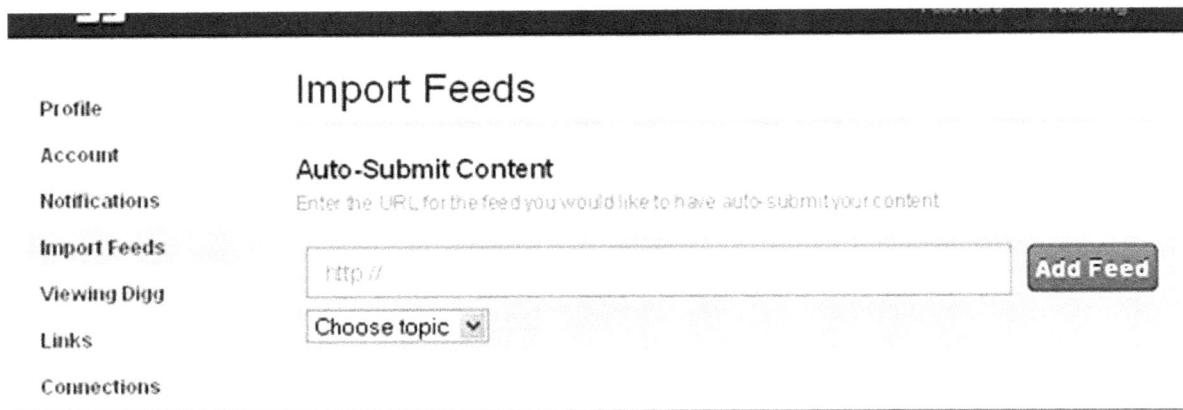

Type in the URL for the feed to your blog. *Note: This is the actual feed address, not just the blog's URL*. Choose a topic and click *"Add Feed."*

Building Followers on Digg

Since your followers have a direct impact on how and if your submitted stories move up the ranks, you'll want to increase your followers as quickly as possible. Fortunately, there are several things you can do to increase your popularity on Digg:

- **Follow Others**—You can't expect people to follow you if you don't follow them. Use your social media accounts as a starting point and then seek out other Diggers that you'd like to follow.

- **Submit Stories Other Than Your Own**—From a marketer's perspective, the whole point of social media and social bookmarking is to generate traffic to their primary website. But if the only links you're promoting are your own, you won't get very far. Instead, make sure to balance links to your work with links to other websites and resources.

- **Say Something**—Digg gives you the ability to leave comments on submitted stories. The more you comment (and digg other stories), the more people you'll have the opportunity to connect with.

- **Complete Your Profile**—Upload a picture, add links to your website or blog, and include links to your social media profiles as well. Make your Digg profile as interesting and complete as possible.

- **Show People You're On Digg**—Include a link to your Digg profile on your website and social media profiles to make it easier for people to follow you.

Also remember to post often. The more active you are on the site, the easier it is to attract new followers.

Mixx (PR8)

Mixx works much like Digg in that your bookmarking links can be submitted to a variety of categories, including: news, business and community.

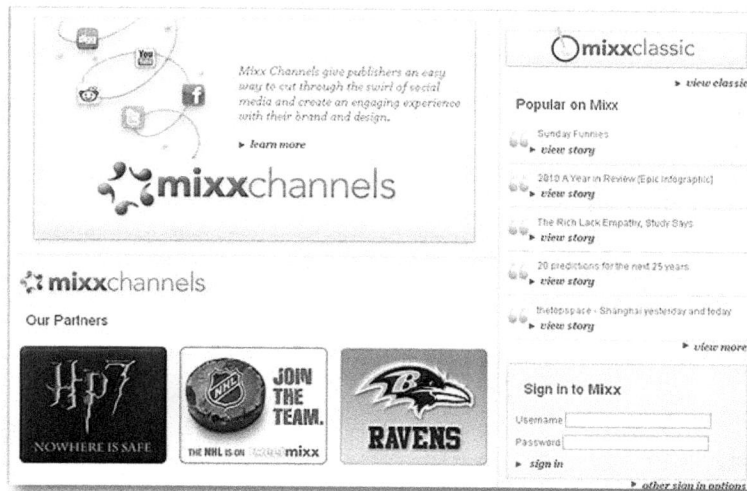

To register, click the *"Sign Up Now"* button just below the login screen. You'll have the option of registering with your email address or using your login from another account such as Facebook, Google or Yahoo!.

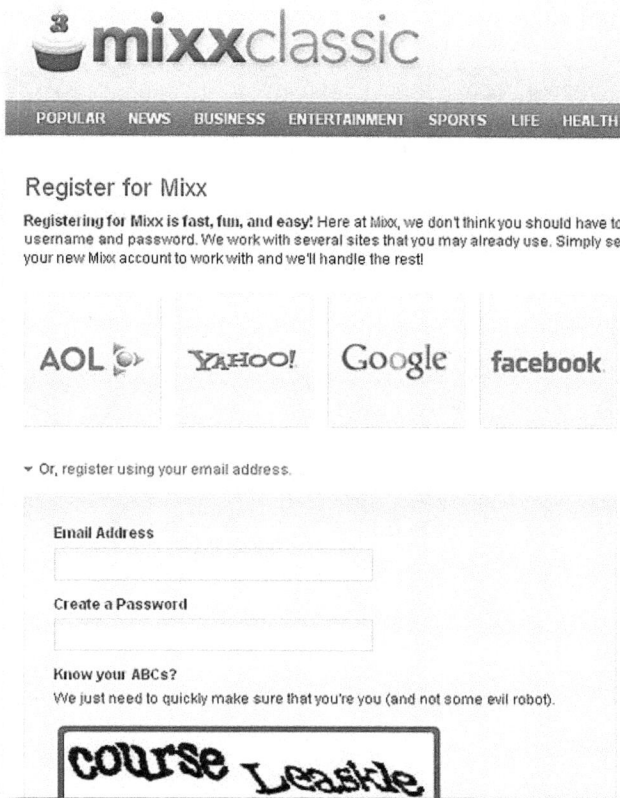

Once you've registered, you'll be prompted to set up your basic profile. In addition to uploading a profile image, you can also choose your User ID, select keywords that represent your interests and include your zip code so that Mixx can match your account with local events and updates. Click *"Submit"* and then check your email inbox. You'll need to validate your account by clicking a link Mixx will send you via email.

Once you've completed your profile, you'll see the main Mixx homepage. To submit a link, click the *"Submit a Link"* button in the top right-hand corner of the page:

Type in the URL of the link you want to submit, verify the format of your link (story, photo or video) and click "*Continue*":

Mixx will pull some basic information over about your link, such as Title and Description, but you'll have the chance to edit this information if you like. You can also add tags (keywords) and a comment as well. Choose your categories, complete the CAPTCHA code on the right-hand side, and click *"Submit"* when you're satisfied with the content.

To vote on other stories (and increase your activity level on Mixx), click the *"Your Mixx"* link. This shows you stories based upon the interests you selected when you set up your profile:

You can also browse stories to vote on by clicking any of the tabs across the top—Popular, News, Business, etc. Once you've found a story that interests you, click the title:

From here, you can cast your Mixx vote by clicking the arrow below the story's ranking on the left hand side (in this case, the number *"12"*) or you can post it to your Facebook account and/or tweet about it on Twitter. You can also click the title again if you'd like to read the full story first.

To see stories that you've submitted as well as those you've voted on, *"favorited"* or commented, click the *"Profile"* link under the *"Your Mixx"* tab.

StumbleUpon (PR8)

StumbleUpon (SU) gives you the ability to like or dislike a site using a *"Thumbs Up"* and *"Thumbs Down"* rating system. If you *"thumbs up"* a site, StumbleUpon uses that vote to refine the pages it shows you in the future. Likewise, if you vote something down, StumbleUpon knows not to recommend similar pages. This is important because your goal here is to build community and SU uses your votes to help connect you with like-minded *"friends,"* so, be sure to vote down as well as up so that SU can refine your results.

In addition to your rating activity, StumbleUpon also uses data from friends (both those you follow and those that follow you) as well as data from similar users with similar (marked) interests. What this means is that the more information you give SU about your interests and your likes, the more selective your SU account will become.

To get started stumbling, you'll need to set up a free account:

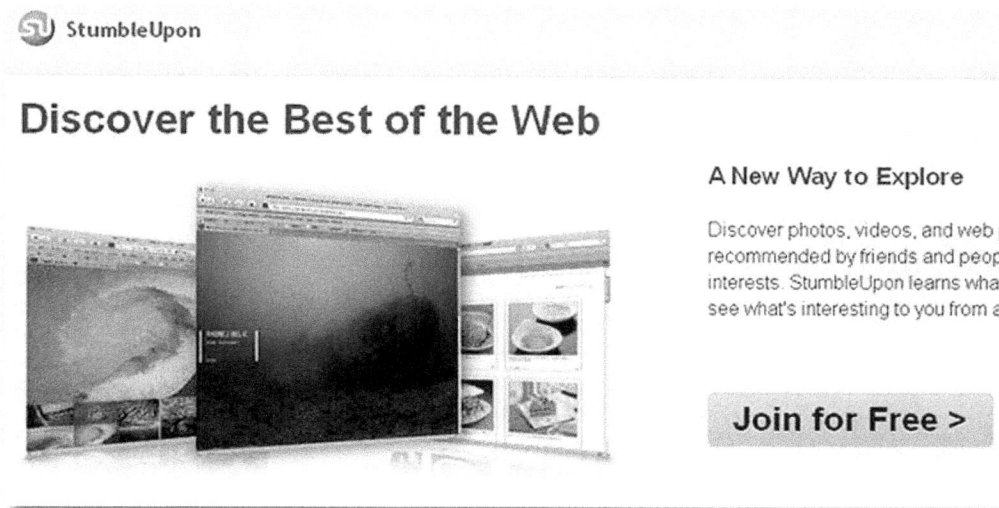

Click the *"Join for Free"* button and enter the requested information. If you don't like the default user name SU creates, you can change it here. Our recommendation: use a user name that reflects your firm (ours is BestLglPractices, for example).

Your email:

Your personal information will never be shared.

Username:

Create Password:

Your Gender:

○ Male ○ Female
(This helps us find the best sites for you)

Your Birthday:
Month ⌄ Day ⌄ Year ⌄

Get Started >

When you click *"Get Started,"* you'll be asked to complete a CAPTCHA and then you'll be taken to a page that allows you to import connections from other social media accounts. You can import these contacts or skip this step. You'll then see a screen that allows you to choose categories that represent your interests.

StumbleUpon

Hi BestLglPractices · Visitors (C

Success! Your account has been created

Now select several topics to start stumbling... (5+ recommended)

Filter by keywords:

Selected 5	☐ Action Movies	☑ Internet	**Start Stumbling >**
Popular Topics (60)	☐ Aging	☑ Internet Tools	
Commerce (31)	☐ Alternative Energy	☐ Ipod	**Suggested Topics**
Arts/History (38)	☐ Alternative Health	☐ Men's Issues	
Society (56)	☐ American History	☐ Movies	☐ Firefox
Sci/Tech (59)			☐ Bizarre/Oddities

When you're done, click *"Start Stumbling"* and SU will take you to pages it thinks you might be interested in.

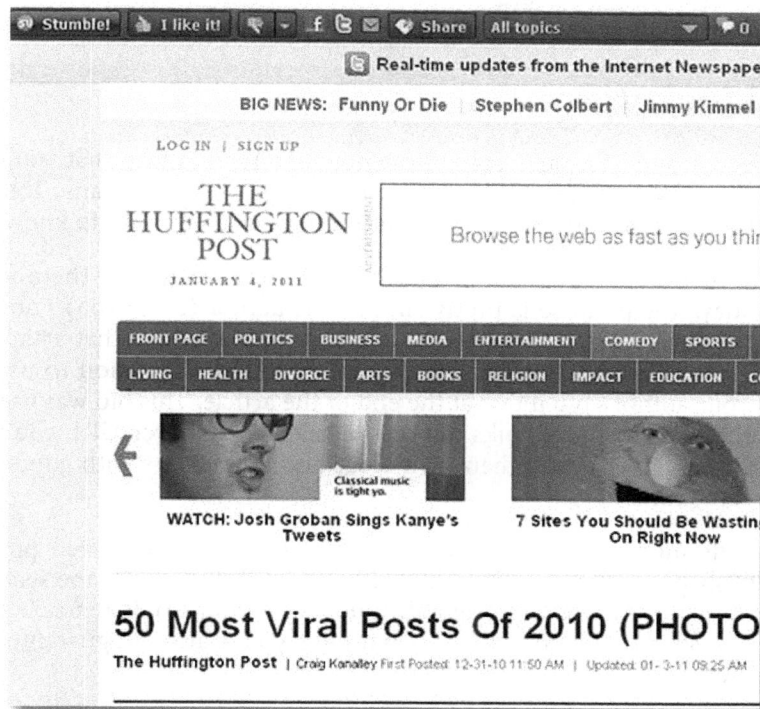

You'll notice the SU toolbar across the top. This is how you rate a particular SU site:

- **Thumbs Up** *"I Like It"*—include this type of content in your results.
- **Thumbs Down**—don't include in your results.
- **Stumble!**—switch to a new page without rating.
    - Social media buttons—email the link or share on Facebook and Twitter.
    - Share—share the link with other stumblers from your Gmail, Yahoo Mail or Hotmail contact list.
    - All topics—go to your dashboard to select a topic you'd like to browse.
    - Quote box/#—if others have reviewed this site, you can see their comments here.
    - # favorites—review your favorites.

How to Find People to Follow on StumbleUpon

There are a couple of different ways to follow someone on SU. First, you can import your own email and social media contacts that use SU. You can also click the user's ID (next to the Quote box/# icon when you're stumbling) to see his/her SU home page. There's a big green *"Follow"* button on this page that allows you to follow this particular user. If you find content on SU that you like, this is an easy way to follow users with similar interests.

11.2: Article Marketing

Article marketing has been around for quite some time now—so long in fact, that you'll find many marketers who say the practice is dead. We're going to disagree. It's not dead, by any means. There are still ways to use article marketing to your advantage and reap the benefits it offers. You just have to know what you're doing.

In the beginning, article marketing consisted primarily of article directories. There are multitudes of these directories still in use today. The premise is pretty simple: the author (that's you) publishes an article on the directory site and then the site's users (other website owners) can download that article for use on their own website, ezine, blog, or newsletter and they can do this for free. The stipulation to using this free content is that the user must keep your author's bio intact at the end of the article. This bio was basically a paragraph that identified you as the author and included links to your website or whatever URL you'd like to promote. This allows the directory site's users to populate their own blogs and newsletters with good content while building backlinks for you at the same time.

The first problem with this method is the duplicate content issue we mentioned previously. If a thousand users download and publish your article on their site, only two will show up in the search results. If these two sites outrank the article directory where you originally published, your original article may not get ranked. In addition, if you've also published the same article on your blog, you're also not guaranteed one of the two spots by being the original author.

The second problem is that, because of the duplicate content, Google devalues the backlinks. That's not to say that you don't still get some credit, just that Google recognizes the backlinks for what they are and gives them less value.

The third problem is that not all article directories are equal. Some, quite frankly, are junk, with no guidelines to regulate the quality of the articles that they take in. As a result, the pagerank can vary greatly from one directory to another. So if you're looking for PR juice via backlinks, you have to be careful about where you're posting your articles.

All that said, there are still a few directories out there worth using and with respectable pagerank to boot.

Ezine Articles

Ezine Articles has a pagerank of six. It is considered to be one of the top article directories on the web. Authors must follow strict submission guidelines for inclusion and all articles must be original, meaning that you can't publish the same article elsewhere on the web. Submissions are reviewed by editors before going live on the site, a process that can take up to seven days. Articles with excessive keywords, poor grammar, or that contain purely advertorial copy will not be accepted. To sign up for Ezine Articles, go to **EzineArticles.com** and click the *"Join Now"* link in the top right-hand corner. To see submission guidelines, go to **ezinearticles.com**/editorial-guidelines.html.

Article Dashboard

Article Dashboard has a pagerank of five and also requires original content. You can include up to three links in the author's resource box, but none are allowed in the body of the article. Articles that do not meet basic guidelines regarding grammar and spelling will be rejected. To register, go to **articledashboard.com** and click the *"Sign Up"* link under Join Our Community.

Other directories include: **GoArticles.com, ArticleBase.com, ArticlesFactory.com** and **ArticleCity.com.** Just be sure to double check the pagerank for these sites and also verify that the backlinks you'll get from the site are *"dofollow."* See Chapter 10 for an explanation about nofollow and dofollow links.

In addition to these directories, you can also add content sharing sites to your article marketing strategy. Here's where your article marketing can really pay off.

Content sharing sites, such as: Squidoo, Associated Content and Hub Pages don't allow their users to republish your content. Instead, you own the *"hub"* of content you produce and it's all grouped together under your profile. That means that if someone is looking for information on wills and trusts, they'll not only see the article you wrote on do-it-yourself wills, but they'll also see links to other articles as well, both yours and articles belonging to other site members. It's like other members are contributing to a site when they visit the content sharing site. It's kind of like having your own blog on a much larger site.

These sites still offer the benefit of backlinks and many have impressive page rankings to boot. Clearly, this is a much better option than the old article directories.

To benefit from content sharing sites, you'll need to create a profile at the ones you want to use. We recommend: Squidoo, Hub Pages, Associated Content and Helium as a good starting point. But, we've included a more comprehensive list in the Appendices if you want to expand on your efforts.

Squidoo

Squidoo is a popular content sharing site that offers a number of unique features to its authors. In addition to posting your content, you can also include a list of related books at **Amazon.com** (using your own affiliate link no less), surveys and polls, a collection of related images from Flickr and even a top ten list that your readers can adjust with their votes.

This interactivity is what gives Squidoo its appeal. So, try to use at least some of the widgets when creating your pages.

To sign up with Squidoo, go to **squidoo.com** and click the *"Sign Up"* link in the top right-hand corner (or the big *"Sign Up Now!"* button). Complete the short registration form and click *"Join Now!"*

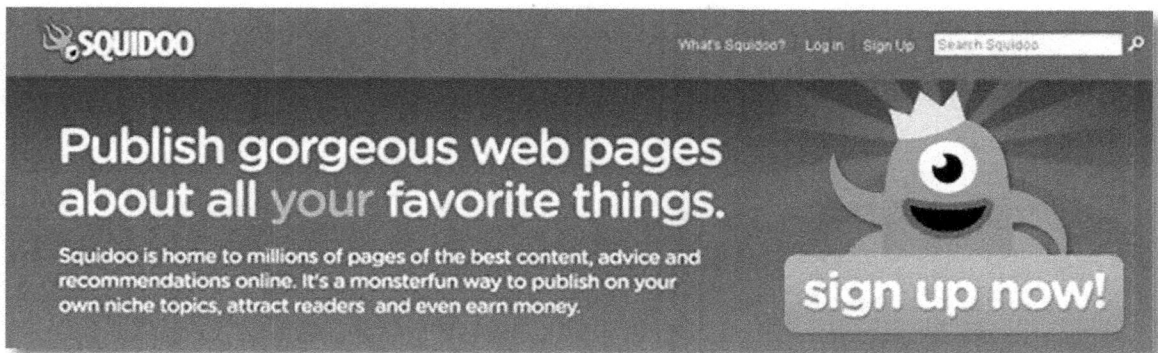

Once you've signed up, you'll be taken to your dashboard, where you can see everything that has to do with your account. To complete your profile, just click on the Profile tab and then click *"Change"* for any information you'd like to edit. Add your website's URL, a profile image, links to your social media accounts and a short bio about your firm.

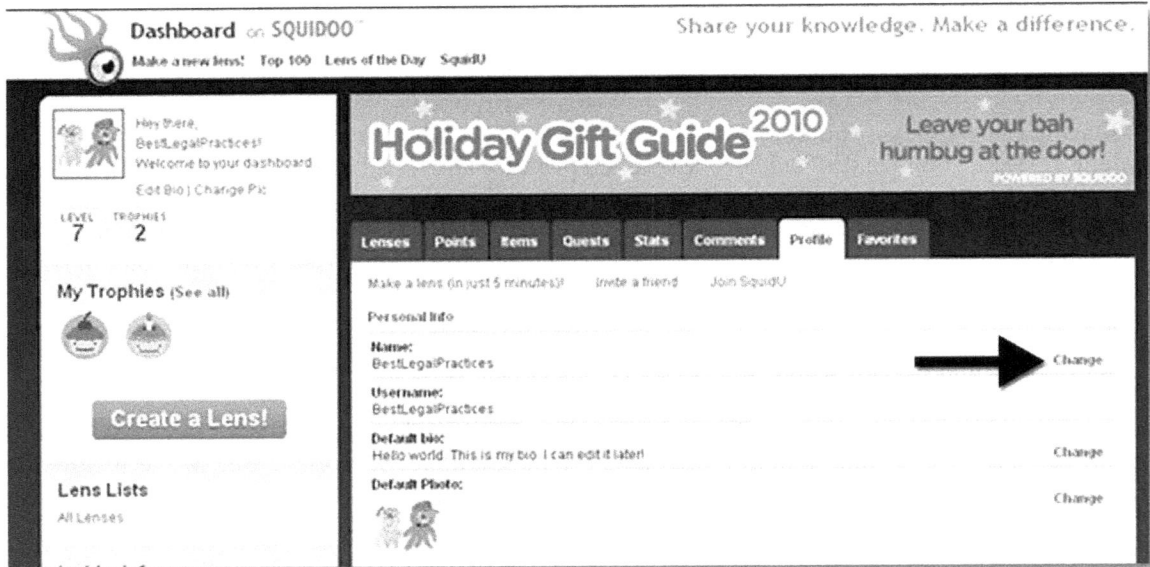

The next thing you'll want to do is set up your first page—called a *"lens."* To do this, you'll need to choose a topic.

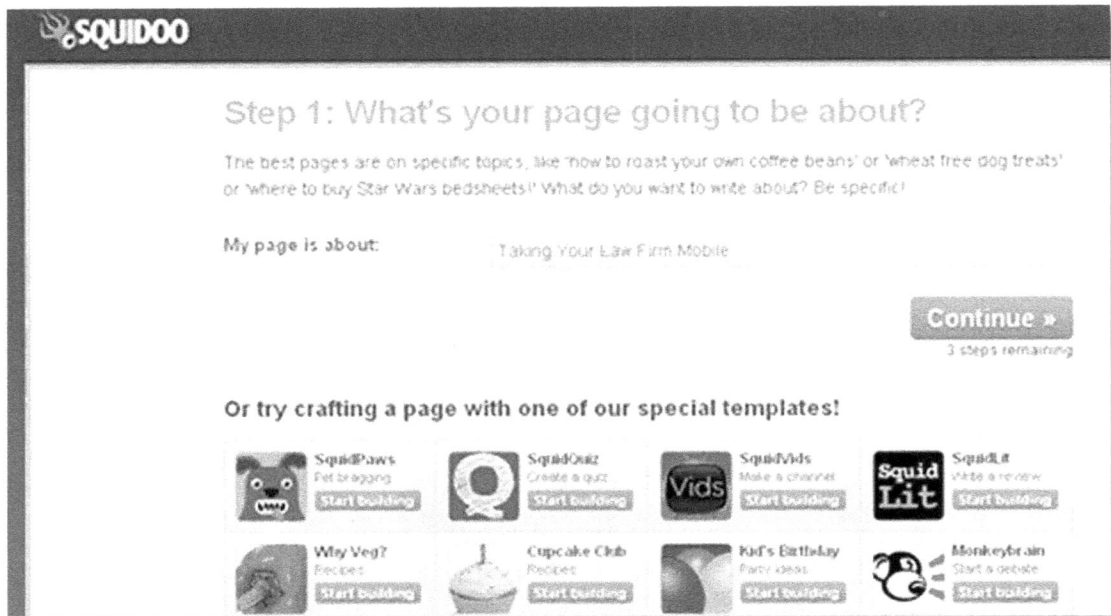

If you've already written an article, then use that. Just be sure to be specific. A single lens won't be about the death penalty, for example, but rather about the debate over whether the death penalty actual acts as a deterrent. The more specific you make each lens, the more lenses you can create on a given topic.

To give you an idea of what your law firm can do with Squidoo, here's the Squidoo profile for Larry Parman, one of our members and a participant in the SEO/SM program we offer at the Academy:

Hub Pages

Like Squidoo, Hub Pages gives you the ability to add extra, interactive modules to your individual pages. To sign up, go to **HubPages.com** and click the *"Join"* link in the top, right-hand corner.

Where Squidoo uses *"lenses,"* HubPages allows you to create *"hubs."* Each hub is a single article, although you can create multiple articles on related topics and tie them all into blog posts, free reports, and other materials you've created. (Yes, this is what you want to do!)

After you've created your account, you'll have the opportunity to select interests that will customize your HubPages experience. Then, you'll be asked if you want to import contacts from your email software. You can skip this step if you like.

Once you've customized your account, you'll see a page that allows you to do several things: create a new hub, upload a profile image and setup your profile.

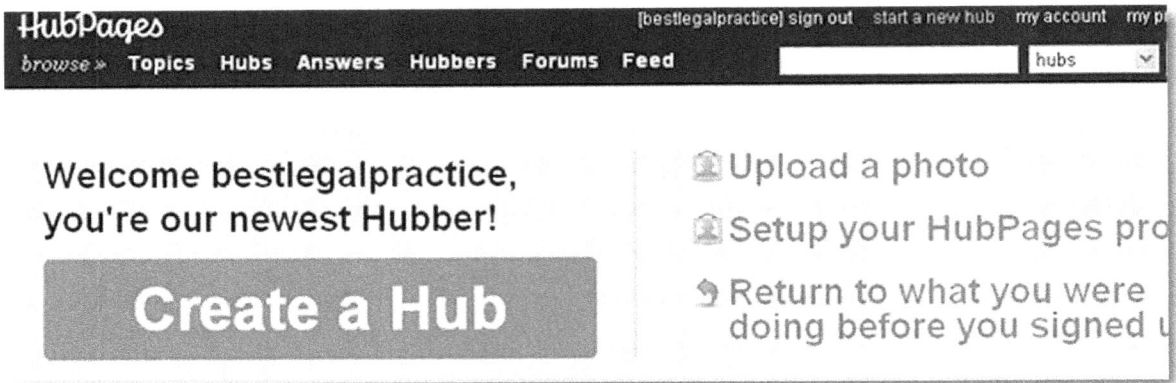

From the profile screen, you can change your password, add a short biography about your firm, add or change your profile image, and choose when you would like to be notified about activity on your hubs.

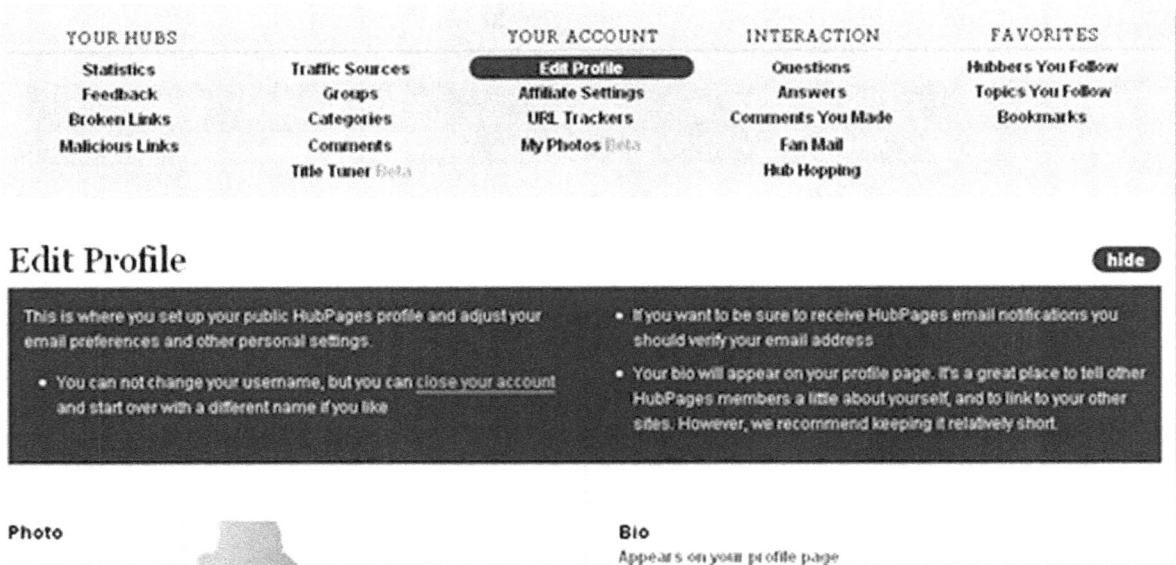

To start a new hub, click the *"Start a New Hub"* link in the top right-hand corner. From here, you can do the following:

- **Title**—Give your hub a title, 120 characters or less. Use something that's catchy and keyword-rich, such as *Five Things Your Estate Plan Should Have* or *How to Settle Your Divorce before It Gets Ugly.*

- **Web Address**—You'll notice as you start typing your title, the web address fills in below, using dashes between the words. You can leave the address as is, or edit if you have something else in mind. Just be sure to make the URL as keyword-rich as possible.

- **Topic**—Select the category that best fits your article. If you're not sure, you can search for appropriate categories by entering your title in the search tab.

- **Layout**—Your layout can be modified as you go, adding, deleting and moving modules as needed. So, just select a layout to serve as a starting point.

- **Tags**—This is basically the same as assigning keywords to your article. HubPages makes some suggestions or you can type in your own.

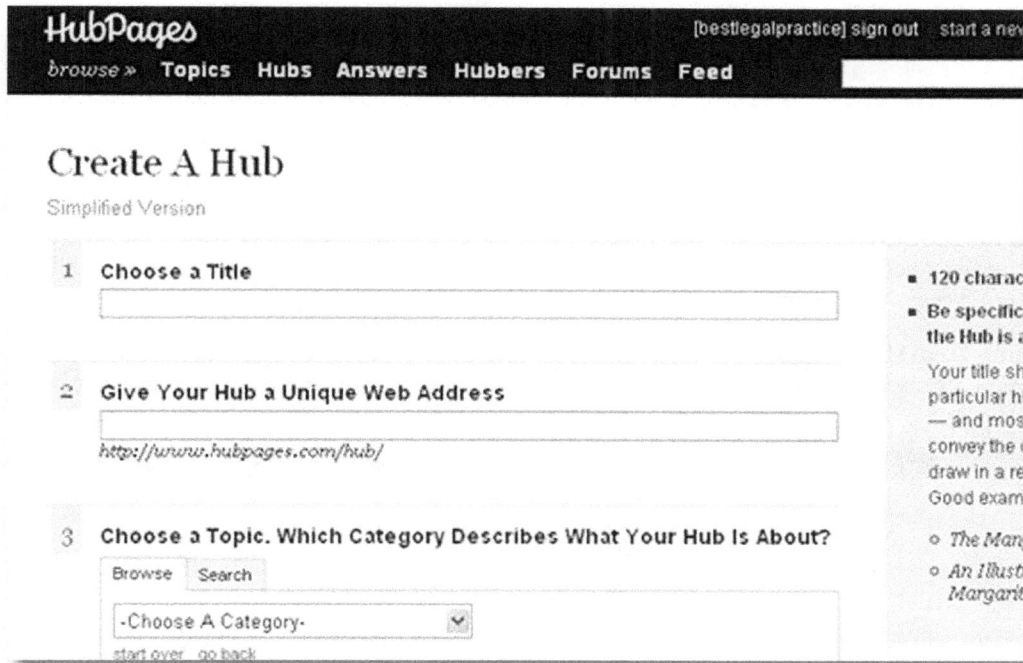

There are two ways to measure your popularity and success on HubPages. The first is to look at your overall Hubber Score. This score is a number ranging between one and a hundred (with 100 being the best). It is shown in the lower right-hand corner of your profile image. Once you start publishing hubs, you'll also notice a number just after the main title of each hub. This HubScore reflects how popular this particular article is with HubPages' visitors. And lastly, you'll see something called HubKarma when you're logged into your profile. HubKarma reflects how often you link to other hubs and to topic pages within your own hubs. Both HubKarma and individual HubScores influence your Hubber Score total.

Here's an example of a HubPages portal created for one of our Academy members:

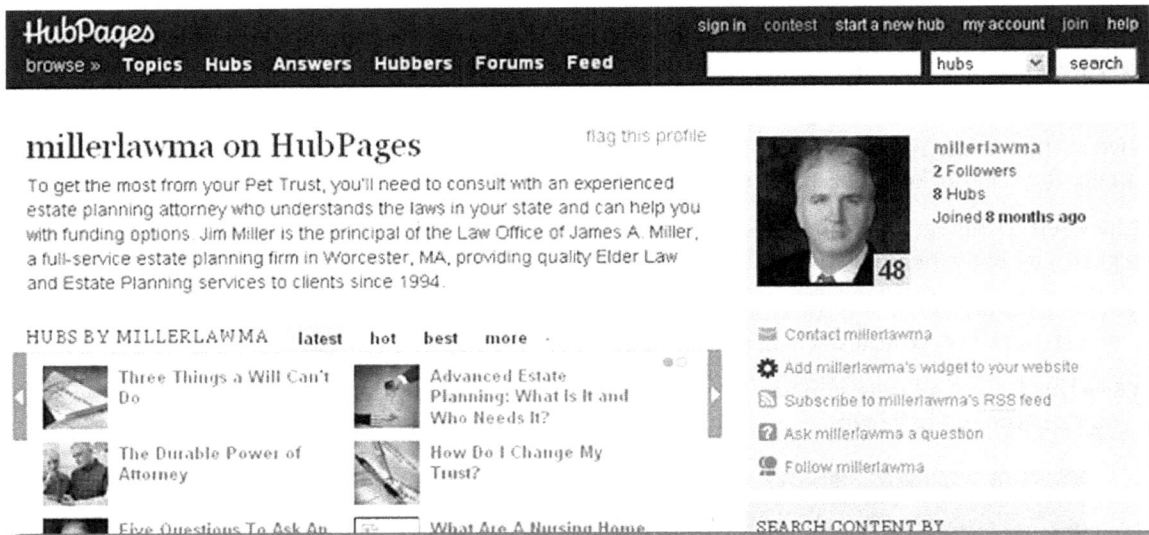

Helium

Although Helium doesn't offer the same number of individual interactive features that you get from Squidoo and Hub Pages, the site does encourage its readers to vote and comment on the articles and it also allows you to earn money through a revenue sharing program.

To join Helium, go to **Helium.com** and click the *"Join"* link in the top, right-hand corner. You'll then be taken to a sign up screen where you can choose what type of account to create. (You want *"Writer."*) Set up your profile.

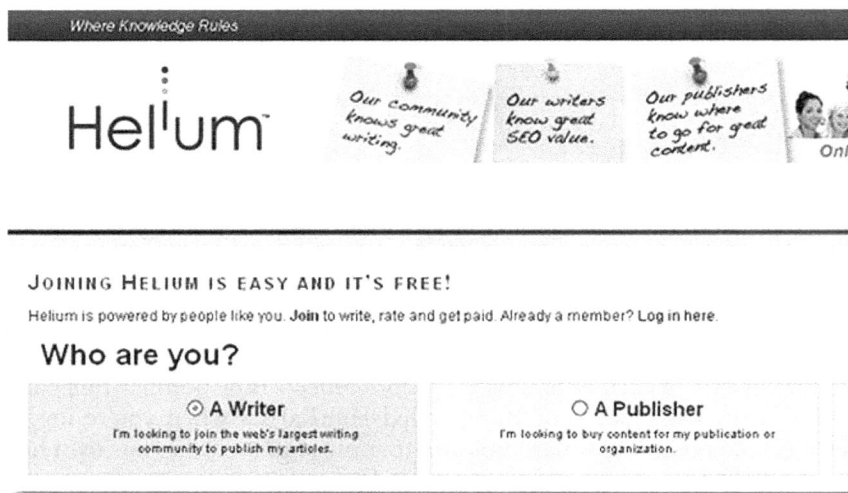

After you've completed the registration process, you can fine tune your profile and start writing your articles.

Click on the *"My Helium"* link in the top right-hand corner.

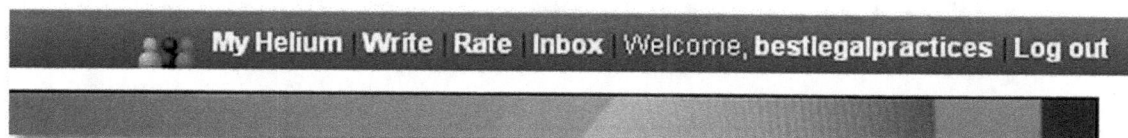

This will take you to your dashboard, where you can view your stats, see recent Helium news, and access all the various features and options for your account.

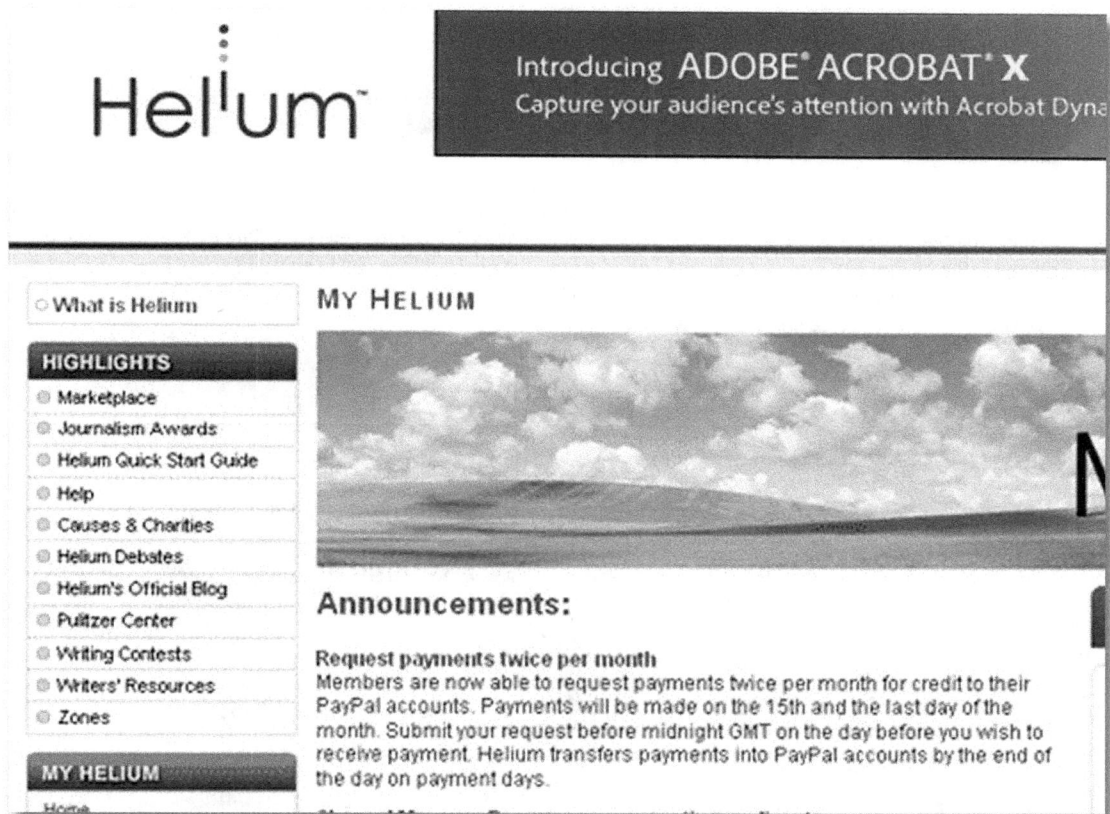

On the left-hand side, you'll see a navigation menu that says *"My Helium."* From here, you can modify your profile, view your articles, how-to guides, earnings and more. How-To Guides are step-by-step guides that you can create in addition to standard articles. To create a how-to guide, you'll need to first join the How To Guide Community by clicking on the *"My How To Guides"* link.

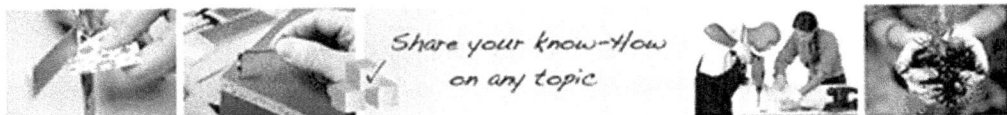

:BETA

Hel'um™

$24.95 a month when you bundle* CenturyLink™ High-S
Real savings don't change

WINTER   SPRING   SU

Share your know-How on any topic

Become a member of our How To Guide community

Do you have professional or personal knowledge on specific subjects that you would like to share? Helium's How To Guide creators are people that are passionate about sharing knowledge with others in an easy to follow step-by-step guide that anyone can understand.

How To Guides are one of Helium's promising new content-building products that are created and leveraged by a collaborative community. Members of the How To Guide community follow guidelines to outline a set of instructions that help people to accomplish common everyday tasks and assignments. Once a How To Guide is approved, the community acts as editors to help increase the quality of the Guide by fixing errors they may find. This empowers the community members to increase the value of each Guide which ultimately enhances the quality of the How To Guide product overall.

Join Now

WRITE NOW!
Wherever you see this icon: Write Now ✎ click on it to add your article to that title.

Find a Title

Looking for inspiration?

Search on subjects you find interesting and find the titles looking for your articles!

Go to Title Finder

Suggest a Title

We've got thousands of titles on Helium. But if the one you want isn't here yet, here's where to add a new one.

Go to Suggest a Title

Earn more in Marketpl

Helium's publishin Marketplace are pa for selected article out the latest Marke

Go to Ma

Create a HowTo Guide

Now you can share what you know on Helium with pictures, illustrations

Write News and Journalism

Whether you want to cover your corner of the globe or report on world

Get in on Debates

Helium brings civil discourse back to the Internet. Pick your side, write your

Enter our Writing Cor

When you write for you compete with

To write new content (articles, guides, etc.) choose the *"Write"* link in the top, right-hand navigation menu.

The Helium system works a little differently than some of the other content-sharing sites you'll work with. Instead of just creating an article from scratch each time, they want you to match your content to their categories and other content available on the site. To get started on a new article for example, you can click the *"Find a Title"* block.

TITLE FINDER

Find Results

With all these words

Exact phrase

With at least one of the words

Number of results per page:

50 results

Type of Title

⊙ All   ○ Debate only

Sort by:

By Relevancy

Show titles

⊙ All
○ Marketplace only
○ Empty only

For titles only in channel:

All Channels

You have selected All Channels

Search

All the suggested titles are keyword-rich, so use the search function to find titles within your practice area. A search on *"divorce"* for example, returned the following results:

TITLE FINDER RESULTS

Back to Title Finder

Results 1 - 50 of 710 'divorce'

Sort by:
By Relevancy

Title	# of articles	Title creation date
How to get *divorced* in the US	7 articles	01/10/08
How to file for *divorce* without contacting your spouse	2 articles	04/07/06
Divorce 101	16 articles	01/31/09
How to have a friendly *divorce*	5 articles	09/04/07
How to have a friendly *divorce*	2 articles	02/21/10
How to tell your wife you want a *divorce*	1 article	01/19/10
How to tell your husband you want a *divorce*	2 articles	01/19/10

You'll notice that this system provides you with the number of articles already written for each title. This lets you see which topics have been covered several times vs. those that haven't been addressed.

If you see a title you want, click it:

| Children & Divorce | Coping with Divorce | **Divorce & Legal Issues** | Divorce Psychology |

Get a **Widget** for this title

How to file for divorce without contacting

Top Article | All 2 Articles

1 of 2 — **by Roz Romero**
There are many reasons people choose to leave their marriages. Sometimes severe financial difficulties squeeze the love right out of the hearts of once blissful couples. Yes, marital money woes lead many down the not so yellow
read more

2 of 2 — **by Susan Terry**
The United States Constitution provides all citizens Due Process of Law. In essence, this means you are prohibited from suing someone in court without notifying that person that he or she had better do

This screen shows you the articles already written under your chosen title. If you want to write an article using this title, just click the small pencil icon above the article listings:

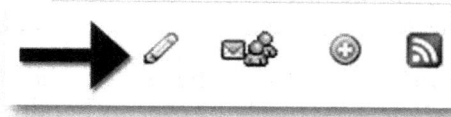

If you don't want to use this title, you can click your browser's back button to see other related article titles or browse titles under different tabs:

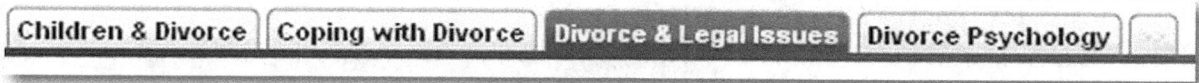

Still don't see what you want? Then you can suggest your own title by going back to the main *"Write"* menu and choosing *"Suggest A Title."* Here's what the Helium profile page looks like once you get going. For this example, we used the profile for Parman & Easterday, a member firm of the Academy.

Associated Content / Yahoo! Contributor Network

Associated Content (AC) is a popular content sharing site with a journalistic feel. It was recently purchased by Yahoo! and is now called the Yahoo! Contributor Network. This network consists of the original AC site as well as Yahoo's many other content sites, such as: OMG!, Yahoo! News, Yahoo! Local, Shine and Yahoo! Finance.

That means that if your articles are well-received with the readers, they have the potential for getting some serious exposure within the Yahoo! network. We're talking over six hundred million visitors per month here, so this one is well worth your time.

To join, go to **AssociatedContent.com** and click the *"Sign up/Publish"* link in the top, right-hand corner.

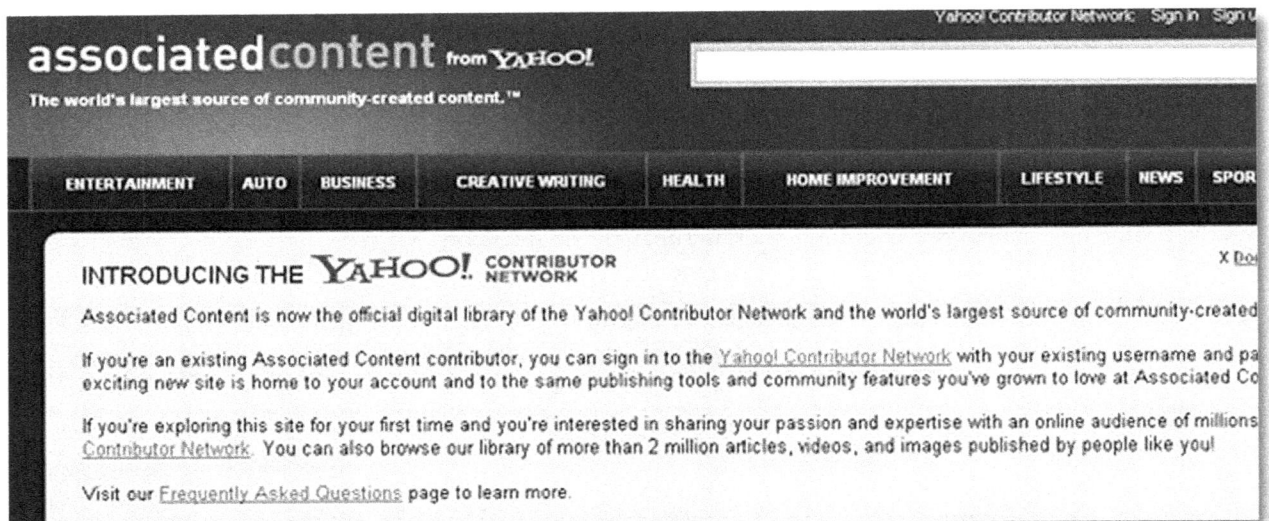

Once you've confirmed your email and activated your account, you'll see your dashboard page:

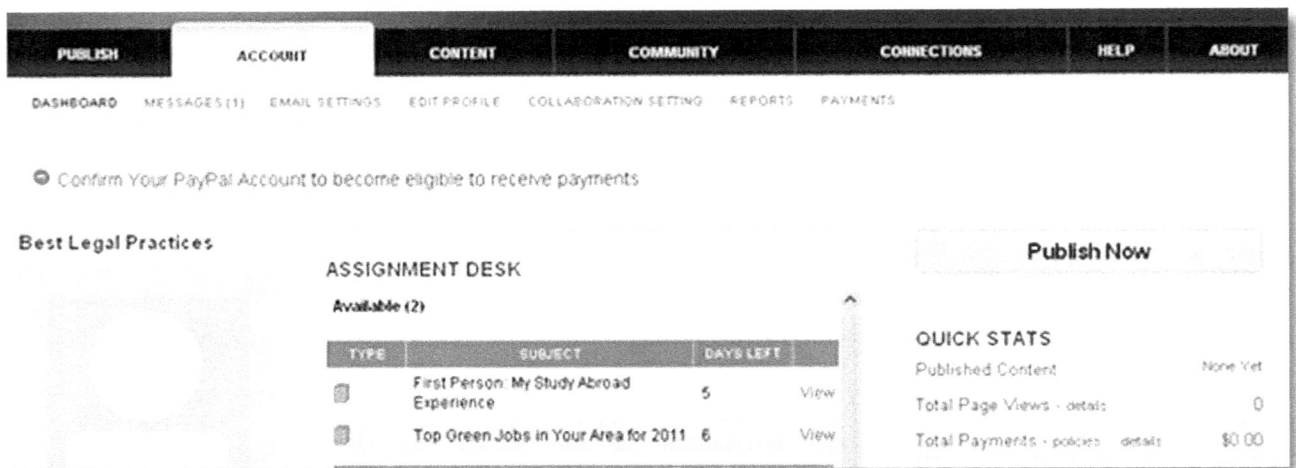

From the dashboard, you can monitor any earnings you've received from publishing, edit your profile, and of course, publish your content.

To write a new article, click the *"Publish"* tab in the top left-hand corner:

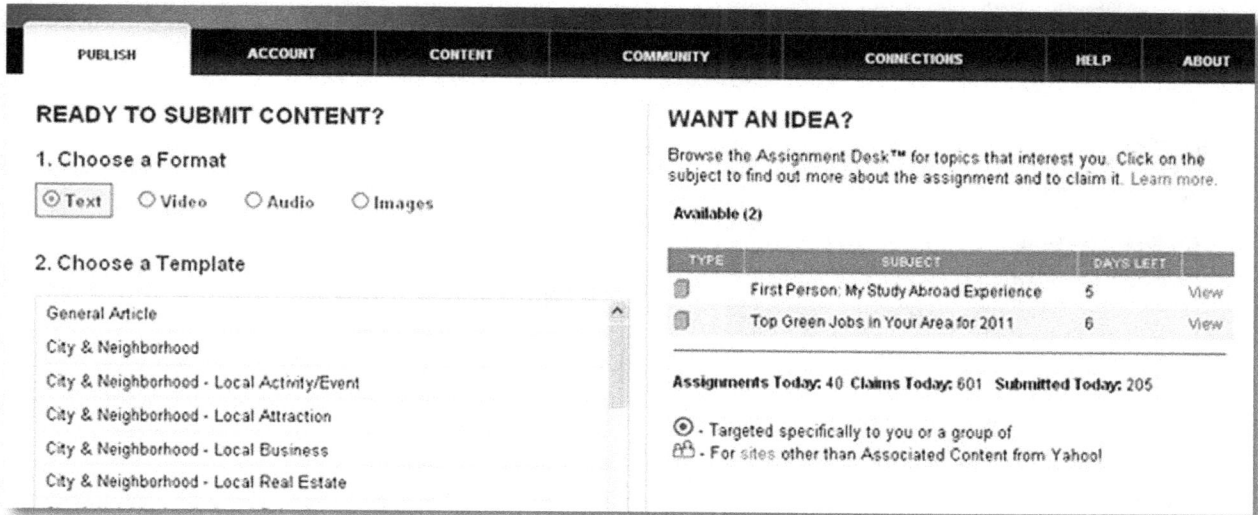

You can create a regular article by choosing *"Text"* format and then *"General Article"* as your template. You'll notice that you also have other templates to choose from, such as News Item, Guide (how-to) and Local Activity/Event. Think about how you could use these additional templates to expand your article submissions and promote your expertise.

On the right-hand side of the screen, you'll see the assignment desk. This is where specific articles have been requested and if you choose to write one, you'll get paid a flat fee for your efforts.

In addition to payments for assigned articles, the Yahoo! Contributor Network also pays you for page views to your content. These payments are made via PayPal, so you'll want to set up a PayPal account if you don't already have one. Just go to **PayPal.com** and choose the *"Business"* account setup.

The Yahoo! Contributor Network works on *"clout,"* similar to HubPages HubScores. New accounts start with a clout rating of one and then increase as they contribute more content and their page views grow. The highest clout rating you can receive is a ten, but once your clout reaches seven, you'll start earning more for your page views. The image below shows what a complete profile looks like on Yahoo! Contributor Network. For this example, we used the profile of Pablo Palomino and Legacy APC, a long-time Academy member.

What Should I Write About?

Working with the same types of topics you'd choose for your blog, create original articles that can be submitted to each of these sites. Note that while you can certainly write about the same topic on multiple sites, each article should be considerably different to avoid the duplicate content penalty. Also, some sites strictly prohibit duplicate content so double-check the guidelines before submitting.

Most of these content-sharing sites also allow you to create in-depth profiles and you'll want to take advantage of this whenever possible. Include images, your website's URL, your social media links and anything else that can help promote your firm.

Once you've published to a content-sharing site, you'll want to make sure to promote your new article on your various social media profiles as well as the bookmarking sites and yes, even your own blog. The idea is to create a continuous circle of information. You publish here, you promote there, and vice versa.

How Often Should I Publish?

This is a question you're going to have to answer on your own. We recommend publishing to at least one content-sharing site each week but that's going to depend upon your schedule and how many of these sites you're including in your marketing campaign.

For example, if you go with the four sites we've mentioned here, you might want to rotate them weekly, allowing you to publish at each site once a month. More sites would require a different schedule—two articles a week for instance—because you don't want to go several months without publishing on a given site.

The good news is that these sites don't require you to publish on a regular schedule at all. That's strictly your decision. So, you can play around with the schedule, and the sites, to see what gives you the best results. Just remember that the more you publish to these sites, the more backlinks you're creating and the more buzz you're generating, all of which supports your position as the go-to law firm on your particular topic.

11.3: Landing Pages and Special Offers

A landing page is a unique kind of web page. Its sole purpose is to drive traffic through a predefined *"funnel"* and generate a specific response from the user.

You could use a landing page to build your mailing list, for example, by inviting people to sign up for the list in exchange for a free report. Landing pages can reside on your domain or they can have a domain all their own. But, their entire existence is to support your primary call to action.

So, let's look at how a landing page is designed:

Landing Page Structure

A typical landing page consists of only one or two pages at most, although we have seen them with three or four, depending upon the options available to the reader. The primary landing page—that is, where the reader *"lands"* when they click your link—can be structured in one of two ways: a short, to-the-point offer, also called a *"squeeze page"* or a longer page with a more *"sales-y"* style of content (aka the sales letter).

Squeeze Page
The squeeze page is designed to exist *"above the fold"* meaning that the reader doesn't have to scroll to see the content. To do this, you might need to design your squeeze page with two columns, using the left column for your promotional text and the right column for your sign up box. You can include graphics with this style of landing page as long as they don't detract from your actual content.

The content itself typically consists of a major headline: **Warning: Your Estate Plan Might Be at Risk!** or **Don't File for Divorce until You've Read This Free Report!** The supporting text would follow below. There are a number of different ways to lay out a squeeze page, but we've found that the following structure will generate the most clicks:

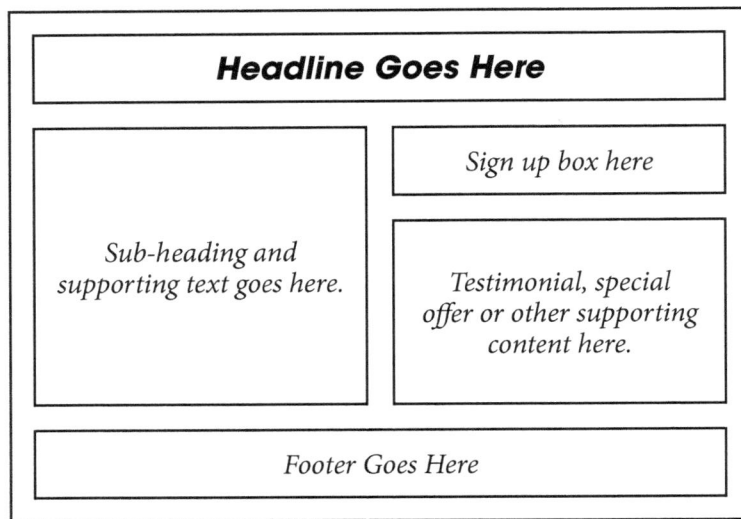

In addition to the content, you'll want to include a thumbnail image of your free report cover or whatever it is that you're offering—place this somewhere around or within the sign up box if at all possible.

Use bullet points and subheadings to break up the text and write in short, powerful blurbs.

Want to see what a squeeze page might look like? Here's one we did on simplifying the probate process for members of our SEO/SM program at the Academy:

Have You Been Named Executor To An Estate...

but don't know where to start?

Finally! Everything You Need To Know About Probate In One Easy-to-Read FREE REPORT!

Losing someone we love is never easy... and the last thing we want to do is worry about probating the estate.

But if you've been named executor, that's exactly what you must do!

Unfortunately, probate can be a very intimidating process. There's taxes to file, creditors to pay and heirs to notify. You'll need to inventory the assets, distribute them correctly and deal with any objections from other family members along the way.

And that's just the easy part.

Sometimes, there's no Will to use as a guide. Sometimes, your loved one has property in another state or worse, the entire estate must be probated in a different state... what do you do then?

Fortunately, probate doesn't have to be a problem and our free report - *Taking the "Problem" Out of Probate* - can give you the information you need to start making some decisions.

Get instant access to this exclusive report! Just enter your name and email address in the box below!

Email :

First Name :

Last Name :

Get Instant Access Now!

In *Taking the "Problem" Out of Probate*, you'll discover everything you need to know to get through the probate process.

- How much does probate cost?
- What does an executor do?
- What property must go through probate?
- How do I handle taxes and creditors?
- What if there is no Will?
- When can a Will be contested?
- What do I do with life insurance policies and pensions?
- What if the probate or the property is in another state?
- **AND SO MUCH MORE!**

In fact, *Taking the "Problem" Out of Probate*, is perhaps one of the most comprehensive guides on the probate process available and you can get your copy ABSOLUTELY FREE!

Just enter your name and email address in the box above to get instant access to

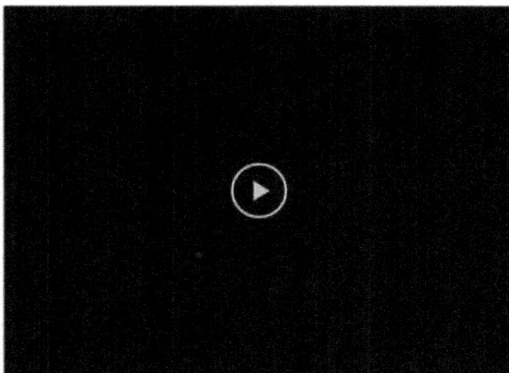

Long Sales Letter

The sales letter is a longer style of landing page. It can range from two pages to eight or even ten. Because of the length, you won't be able to condense all your content above the fold. So, instead of splitting your content into columns, you'll use one main column with boxes inset throughout to highlight important information.

Like the squeeze page, the sales letter starts with a bold headline and usually also includes a smaller, sub-headline right below. So, for example, an estate planning attorney giving away a free report about the importance of having an estate plan might have a primary headline that reads:

Urgent! Your Estate Could Be at Risk!

and a sub-heading that says:

Find out how to protect your assets and those of your loved ones before it's too late!

The style of writing is totally up to you. But, we can tell you from years of experience that the message needs to be a powerful one. These headlines might seem a bit cheesy at first glance, but you want to get your readers' attention. Their estate *could be* at risk after all, and your free report is the first step in finding out how to protect their belongings and those of their family members.

What follows the headlines is a combination of testimonials, bullet points, additional sub-sections and sign up boxes, all of which entice the reader to do whatever it is you're asking him/her to do.

These sub-sections tell a story or address a concern and follow a 1-2-3 kind of process: Here's the problem, wouldn't it be great if...Look! We have the solution.

Each section is designed to support the notion that the reader should buy but because of the length, you'll want to write in short sentences and use bullet points and formatting options (bold, italic, font size) to break up your lengthy sales letter.

These various sections can be further segmented by inserting a horizontal line and/or by creating a single-celled table with a different background color.

For example:

Headline and sub-headline here	
Intro text here	
Sub-headline here	
■ Bullet	■ Bullet
Sign up box here	
Sub-headline here More content	
Sign up box here	

You'll notice that there's more than one sign up box on the sales letter. This is important. Because the sales letter style is so much longer, you'll want to offer your readers the ability to take your offer without having to scroll to the top or bottom of the page. Scrolling may not seem like a big deal to you but it makes a big difference in conversions so space your sign up boxes throughout your sales letter for maximum effect.

You'll also notice that the content of the sales letter does not extend all the way across the page. This is a design strategy that should always be followed. Use a white background (with the exception of the colored section boxes, of course) black text, and a margin setting that utilizes only 70 to 75% of your page width.

You can also include images throughout the copy. In fact, it's more appropriate here than in the squeeze page because you have more space to work with.

The content itself should *"flow,"* meaning that it identifies the issue (your headline), introduces the solution (your firm), and then highlights all of the various benefits the reader will get from taking the action requested (bullet points and sub-sections).

Incidentally, this is where the services of an experienced copywriter would be helpful. There is an art and science to writing persuasive copy.

Landing Page Usage

Okay, so now that you know how to create a landing page, let's talk about how to use it.

Remember, your landing page should have a specific purpose. So, you should have a separate page for each offer you want to promote. Landing pages work best when there's a free giveaway in the bargain. This is another area where having a good writer can be beneficial. A simple free report or an e-book makes a great giveaway item and is usually enough to entice most of your visitors to sign up for your list.

To create this type of marketing material, just look at the articles and blog posts you're creating. Most of those topics could be easily expanded into something more in-depth and turned into a special report or an e-book. You could also use the same content to create a webinar, teleseminar or educational video. In fact, many landing pages feature a video as the *"giveaway."* The reader just has to sign up for your mailing list in order to view the video in its entirety.

In addition to growing your mailing list, you could also use a landing page to generate interest for a new webinar, teleseminar, or on-site seminar at your law firm. A landing page would also work well to solicit appointments for a free consultation, by offering that meeting as a *"bonus"* to your free report giveaway.

You can have as many landing pages as you need. Just be sure to track them separately so that you can tell what's working and what's not. We'll get into tracking your results a little later. So, don't worry if you're not yet sure how to measure your traffic.

Separate Domain or Not?

We mentioned earlier that landing pages can reside on your existing website, **SmithDavisLaw.com**/landing1, for example. But landing pages can also have their own domain name, such as **SanDiegoDivorceAttorneys.com** or **ProtectYourAssets.com**.

So, which one is best?

Both work equally well. If you've purchased extra domain names that target a specific aspect of your practice, you may want to use that for an equally-targeted landing page. In theory, you don't even have to worry about separate hosting since you can still host the actual page on your existing website and just create a 301 Redirect for the separate domain.

The only real issue here is what works best for *you*. If you've got the extra domains, go ahead and use them. If not, don't worry about it. Just promote your landing pages from within your existing site.

11.4: *Videos*

We've mentioned using videos in other areas of this book, so let's address that tool now.

Videos are very possibly the biggest marketing tool you have at your disposal.

Links by Media Type

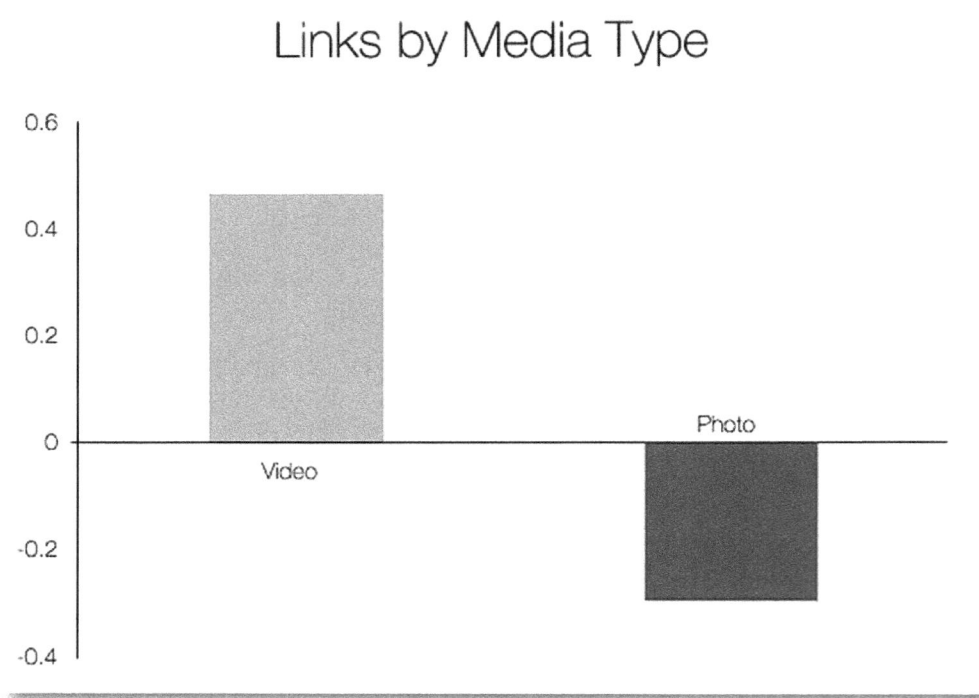

Videos are shared more often than photos (see image above) and consistently receive more page one rankings than plain old text pages. In fact, recent studies from Forrester Research shows that a video is fifty times more likely to wind up on page one. This may change of course, as time goes on. But, at this writing, you're not really marketing if you're not making videos.

---

1   *HubSpot, The Science of Blogging,—www.hubspot.com/the-science-of-blogging/*

Now, there are different types of videos you can create. So, let's talk about those first. Then we'll look at how to use them in your marketing campaign.

Video Styles

There are four basic types of video: live footage, images with voice-over, text with voice-over and slideshow videos with a musical background.

Live Video
While you can certainly do *"live feeds"* for webinars, the live videos we're referring to here are videos where you are the main attraction. These videos can be used for press releases and educational tools. They require only a script, a camera and someone to deliver the content. Live videos can be shot using a web cam or a standard video camera as long as the footage can then be uploaded to your computer.

Images with Voice-Over
Another common style of video you'll see are those where the creator has put together a collection of still images and presentation slides that highlight the important parts of the video's message and illustrate the overall topic. To create these videos, you'll need a video editing program, such as Camtasia, to piece together the slides and images and set the transition timing between them. You'll also need a microphone to record your voice-over.

Text with Voice-Over
Like the images with voice-over, this type of video uses presentation slides to deliver the visual portion of the video, but instead of showing images, the slides simply highlight important parts of the script by allowing the text to *"unfold"* on the screen as the narrator reads. This type of video will also require a video editing program, a presentation program (such as PowerPoint) and a microphone to record your voice-over.

Slideshows with Music
This is a relatively new type of video in the marketing arena but it works great for firm introductions and promotional type videos. In this variation, a montage of images and text unfold on the screen, telling your *"story"* against a musical background. With the right script, images and music, the result can be very powerful.

Now, all these styles can be used to serve a variety of needs. What's more, you can combine the various styles if the project warrants it.

An estate planning attorney for example, could create a live video that explains the difference between a Will and a Trust while a corporate attorney might lay out the reasons to incorporate. In addition, both firms could use a montage-style video as an introduction to their firm or to highlight concerns that are common in their respective practice areas. Slides and images can be used to zero in on one important point. Then, the video could switch back to the live format, allowing the attorney to deliver a powerful *"close."*

And if that doesn't inspire you, here are a few other ideas you could use in video:

- **Answer common questions**—Pick a question, any question, that's asked frequently by new clients and then answer it in a video. One question per video means you've suddenly got tons of material to work with.

- **Tell a story**—People love stories so if there's a way to talk about a set of circumstances without violating any confidentiality and you can make it lively, emotional, or just generally interesting, you've got yourself a video script.

- **Educate**—Tell your potential clients something they didn't know. Explain how the court system works or what they should look for when hiring an attorney in their practice area. Explain terminology or procedures that frequently frustrate new clients. Host a webinar (That's a seminar on the web.) and post the video for public viewing. Teach your audience something and they'll keep coming back for more.

- **Promote Your Giveaway**—We've talked quite a bit about offering free reports and e-books as giveaways. Why not use a video to promote them?

One expert suggests creating ten videos on the questions your prospects and clients ask the most. Then create ten more videos to answer the questions that they should be asking but don't know to ask.

The beauty of video is that it's limited only by your imagination. Just like commercials, videos can be used to: promote your brand, highlight a service, expose a need, connect with your community, and/or introduce your firm or drive traffic to your website.

So, just how do you make a video?

Video Dos and Don'ts

Be Creative
Yes, you can definitely use your articles and even the marketing copy from your firm's brochure to craft a video. But keep it lively and interesting—no boring, monotone video scripts, please.

Short and Simple
Videos are popular because they capture your audience. But if your video is too long or too complex, your audience will start to tune out. If the subject matter calls for a more in-depth explanation, consider breaking your video into *"parts"* so that your viewers can watch the pieces individually.

Include Your URL
You're creating these videos to promote your brand and your services, so it only makes sense to include your website/social media URLs wherever you can. There are programs that allow you to insert a clickable link at the end of your video which is obviously a good idea. Barring that method, you can always include your URL as a *"frame"* at the end of your video and of course, you can post it on your video channel profiles.

Analyze Your Results
To get the most from your marketing efforts, you need to know how your videos are performing. Enter TubeMogul (**TubeMogul.com**)—a free analytic service designed exclusively for video marketers. In addition to simplifying the upload process, you can also see how many times the video was viewed and/or embedded, the search terms that delivered your video in the results, where your audience comes from and more. Go to **TubeMogul.com** to sign up.

Don't Forget SEO
Most video content sites give you the ability to include tags, keywords, categories, descriptions, and titles when uploading a video to their site. Take advantage of this SEO opportunity and make your content keyword-rich.

Submitting Your Videos

Your videos should always be included on your own website. But, there are other places you can submit your works of art as well. YouTube is, of course, the most obvious. But there are plenty of others. We've outlined how to use YouTube here in this chapter. You can see a comprehensive list of video sharing sites in the appendices section of this book.

YouTube
YouTube is, perhaps, the most popular video-sharing site, boasting well over two billion views per day.

Acquired by Google in 2006 for $1.65 billion, YouTube is translated into fifty-one different languages across the globe and accommodates twenty-four hours of video uploaded to its site every minute.

To submit videos to YouTube, you'll first need to create a free account. Go to **YouTube.com** and click the blue *"Create Account"* button:

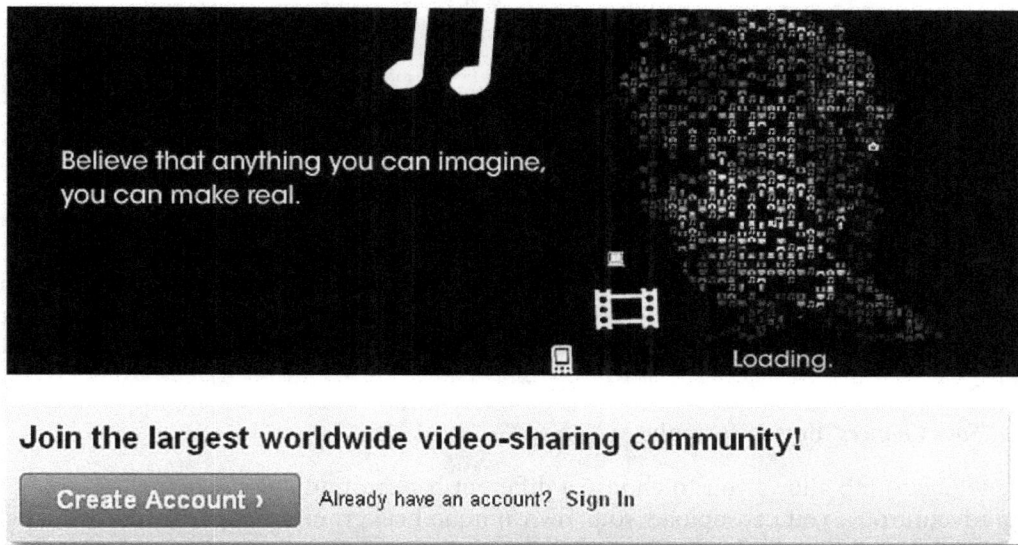

Complete the basic sign up form and you'll be returned to the homepage but this time, you'll be logged in. To access your account, click the small drop-down box in the top right-hand corner that has your username on it:

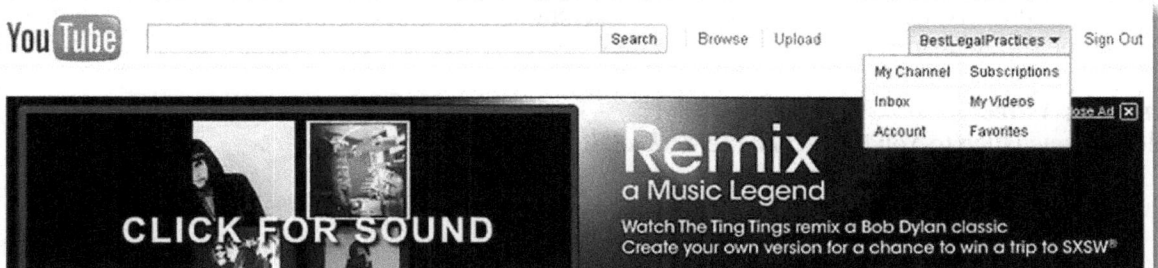

From this drop-down box, you can edit your profile, upload, edit and view your videos, manage your subscriptions and view your messages. Yes, YouTube has an internal email system.

My Channel

This selection will show you an overview of your account: views, subscribers, recent activity, friends, comments, favorites, etc. You'll also notice a box across the top of this page that features tabs: **Post Bulletin, Settings, Themes and Colors, Modules, Videos and Playlists**.

The Post Bulletin button is a relatively new feature but one that you should utilize once your video channel has subscribers. This box allows you to post a message to your subscribers and friends and even include a video link if you have one. This message (and link) is then displayed on your friends' and subscribers' YouTube homepages whenever they log in.

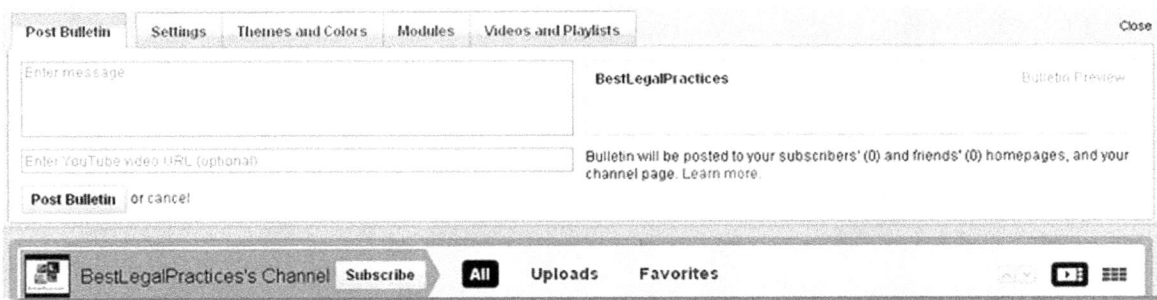

The *"Settings"* tab allows you to fine tune your channel by giving it a title. (We used a variation of our tag line.) Choose your channel type (YouTuber is fine.), decide if you want your channel visible to the public (You do.) and assign channel tags. These tags are essentially the keywords you want to be found under in YouTube, so choose wisely.

Click the blue *"Save Changes"* button to apply.

The *"Themes and Colors"* tab allows you to choose a different background color scheme for your channel or, if you're feeling adventurous, you can upload your own unique background image. To do this, click the *"Show Advanced Options"* link and then click the *"Browse"* button next to the Background Image subheading. Find the image you want to use on your computer, upload and click the *"Save Changes"* button.

The *"Modules"* tab allows you to select which features you'd like to include on your channel's page. The default settings are usually sufficient here.

The *"Videos and Playlists"* tab allows you to choose different layouts for your channel page and designate which video YouTube should feature first.

Uploading a Video

To upload a new video to YouTube, click the *"Upload"* link at the top of the page.

From here, you can upload an existing video or record one from your webcam. Once you've uploaded or created your video, you'll need to give it a title, add tags and a description and select a thumbnail image for YouTube to display in its search results.

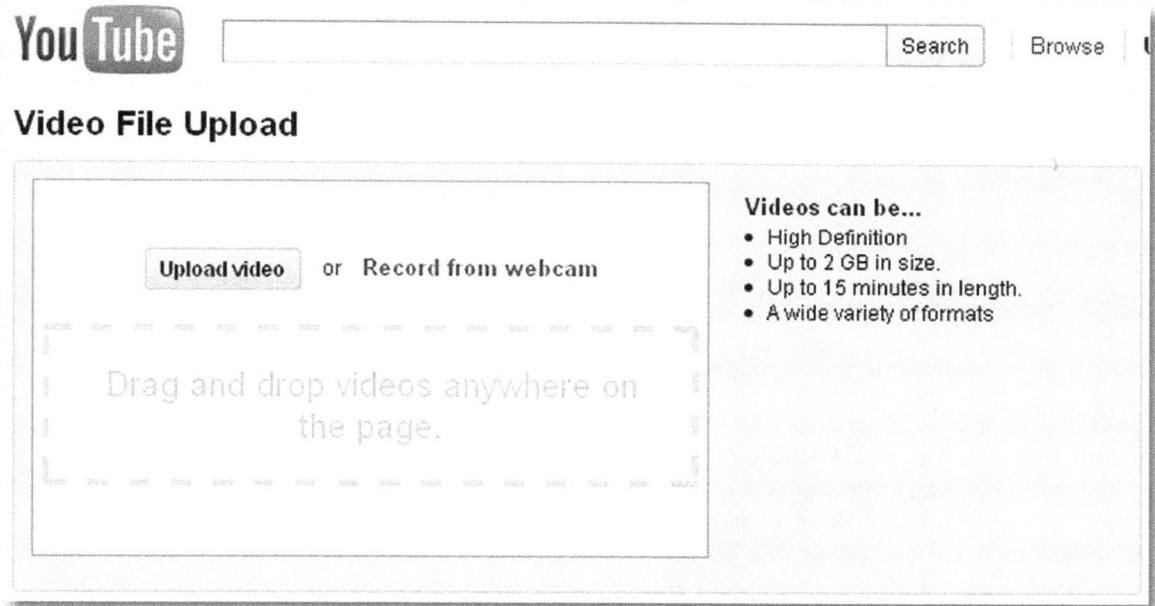

Videos can be up to 2Gigabytes in size and up to fifteen minutes in length. YouTube supports several different video formats, including MPG4, AVI, FLV (Flash) and WMV.

How to Get the Most from Video Marketing

As with any other social media outlet, (Yes, video is considered social media.)you'll want to follow certain guidelines to get the most bang for your video efforts.

The first rule of thumb is to plan your video before you shoot. Storyboard the idea, write out a script and line up any music, images or other extras you plan to include. Practice your video a few times to ensure that your script flows nicely and delivers the message you want to convey.

The second rule of video marketing is to keep your video short. In general, anything over two minutes is considered too long. So if you've got a topic that requires some additional footage, consider breaking it up into multiple videos.

Customize your personal profile on the video sharing sites so that your prospects can find you easily. Upload often and be sure to include links to your videos on everything: your website, your social media profiles and even your email signature.

11.5: Podcasting

Podcasting is the process of creating pre-recorded webcasts (audio files) and posting it on the Internet for download. Most podcasts are created as *"episodes"* to an ongoing series, such as for a radio show or educational course. But podcasting can actually be used for a variety of purposes, including your online marketing campaign.

The great thing about podcasting is that there are plenty of tools out there to make the recording process easy. Once you've created your audio file, getting it out to your potential audience is a snap.

How To Create Your Own Podcast

The first thing you'll need to do is invest in some audio recording equipment. This doesn't need to be anything fancy or expensive. A computer microphone and some audio recording software is really all you need. (We'll discuss some different software options in the next section.)

The most common podcasting file format is MP3. So make sure that your recording software can convert into this format.

Choose your topic. Those articles and blog posts you're writing are a great place to start. Then create a *"script"* that you can read from when you record. Now, that script needs to be engaging—think exciting and entertaining. Remember, there's no video for this type of marketing material. All the audience will have is your voice. So, vary the tone and provide them with content that leaves them wanting more.

Podcast Recording Software

To record your podcast, you can use one of several different software programs:

RecordForAll
This program gives you complete control to record and edit audio files in MP3, WMA and WAV formats. You can import existing audio files, record your own or combine various files using the layer clips function. Go to **RecordForAll.com.**

Audacity

This is an open-source (free) audio editor that includes a fairly decent amount of effects and editing tools considering that it won't cost you a dime. Audacity is available for Windows and Mac. Go to **audacity.sourceforge.net.**

Propaganda

This program was designed specifically with podcasting in mind. In addition to an array of audio recording tools, it also has a built-in publishing system that creates the XML, HTML and MP3 files necessary to make your podcasts downloadable and RSS ready. Go to **makepropaganda.com.**

To get the most from your podcasting efforts, you'll also want to create an RSS feed for your audio files so that they can be picked up by the multitude of directories out there.

To do this, you can use a program with a built-in publishing system such as Propaganda mentioned above, or you can choose a different route. The makers of the RecordForAll software for example, also make FeedForAll, an RSS publishing system designed specifically to turn podcasts into XML-compliant and iTunes ready RSS feeds.

Another alternative is to use a service, such as the one at **Podbean.com.** Their publishing system includes a feed generator, along with management and promotional tools to help you publicize your podcasts.

Podcast Promotion

Once you've created your podcasts, they'll need to be stored somewhere for public consumption, typically your blog or a public podcasting directory. But once you've loaded your podcast, how do you get people to download and listen?

Yes, you guessed it: Podcast directories!

Many of these directories are sorted by topics and categories so you can find the best match for your firm's practice area. In addition to Podbean, there's also **Podcast.net, PodcastCentral.com, Syndic8.com** and **iPodLounge.com** to name a few and let's not forget iTunes, the biggest aggregator of podcasts on the web. We've also included a more comprehensive list of podcasting directories in the appendices section of this book for those who want to get really serious about their podcast promotion.

Getting your podcast publicized is similar to promoting your website. You need the search engines to index the content. One way to do this (in addition to getting listed in the podcast directories), is to subscribe to your own podcast. Since podcasts have an RSS feed, you can use any feed reader to subscribe, meaning you can include your subscription on the big guys such as Yahoo!, Google Reader, Bloglines, and the like. Not all of them will be able to accommodate the audio component of the feed but the process of you subscribing points them in your feed's direction.

Include your podcast feed in the signature line on your emails, promote it on your website and, most certainly, on your social media profiles. Use keyword-rich descriptions and tags when uploading your podcasts to the various directories and, most importantly, publish often.

Podcasts can be downloaded to iPods, iPhones, Smart Phones and so many other gadgets with digital audio playback capabilities. That means that your podcast could be something people listen to on the way to work, while exercising at the gym or during their morning jog. The more you publish, the better chance you have of being found by your target audience and becoming a regular fixture in their audio library.

11.6: *Mobile Marketing*

When smart phones were first introduced, users could surf the web but they had to be willing to scroll sections of a web page individually instead of viewing the entire page at once. This is due to the smaller screen resolution available on a mobile device. Traditional websites just weren't set up to function within such a small space.

But today's mobile user enjoys the convenience of customized sites designed specifically for mobile devices. As a result, these users have a little less patience for scrolling through sections of sites that aren't optimized properly.

And this is a trend that's only going to get stronger.

There are currently over a billion new smart phones and mobile devices being sold each year and while this number continues to climb at a phenomenal rate, the sale of personal computers—yes, we're talking about your faithful desktop!—is starting to slack off.

In fact, the numbers suggest that mobile devices and traditional PCs are currently neck-and-neck. In an April, 2010 presentation at Google, Morgan Stanley analyst Mary Meeker put some firm numbers to what many experts had been suspecting: We had reached a tipping point and current trends suggested that mobile devices would vastly surpass PCs within the next few years.

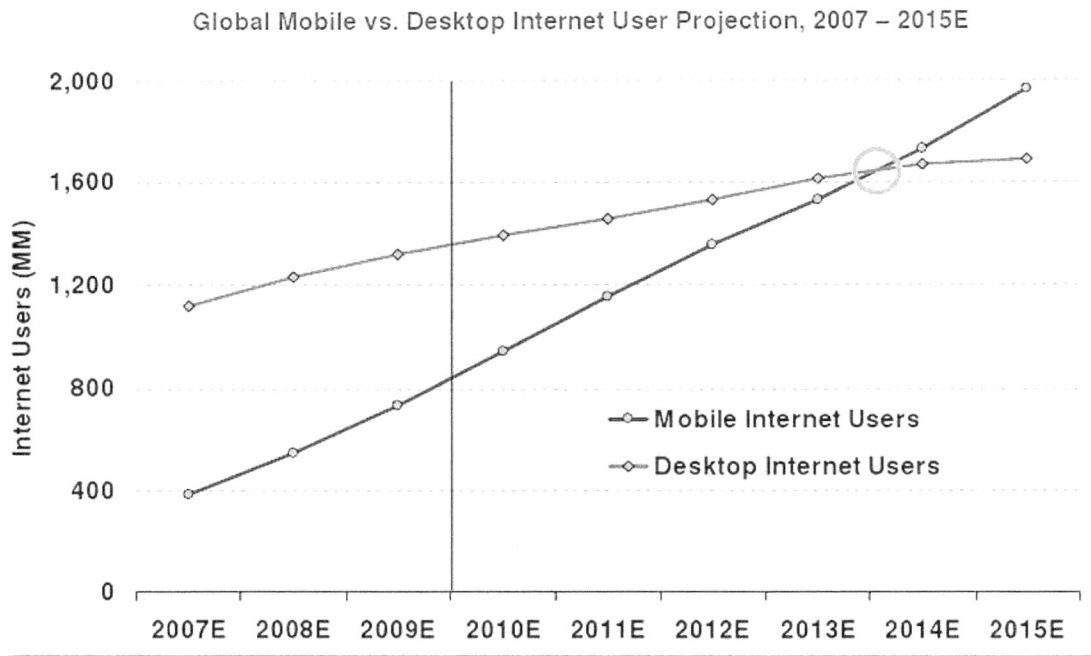

Mobile Users > Desktop Internet Users Within 5 Years

Global Mobile vs. Desktop Internet User Projection, 2007 – 2015E

Apparently, however, these estimations weren't aggressive enough. Mobile devices passed desktop computers in the last quarter of 2010.

Which brings us back to the question at hand: Does your firm really need a mobile version of its website?

Obviously, we're going to say yes, and we've got a few more reasons to back up that answer:

You Want to Be Where Your Clients Are
Let's face it: the whole point of your efforts—heck, even the whole point of this book—is to help your clients find you and find you quickly. As of this writing, half the connections to the Internet are coming from phones and the #1 access method for local information is the mobile browser.

So, if your clients rely on their mobile devices to access information, why wouldn't you want to have a mobile image for your firm?

Search Engine Power
Just as the search engines rank traditional sites according to their usefulness, it also follows that they'll promote .mobi sites first when returning results for mobile devices. Yes, the search engines can tell and if a mobile version of the site is available, they'll prompt the user to be redirected there instead. So, since we're focused on building your brand and establishing your firm as a dominating force in the search engines, doesn't it make sense to maximize your exposure in every way you can?

Protect Your Brand
Just as someone else can purchase the .net version of your .com (and vice-versa), the .mobi version of your domain name can be owned by someone else. Do you really want potential clients finding a different law firm when they're searching for you on their mobile device?

It's Not As Hard As You Think
Going mobile isn't the big ordeal that most people think it is. You can, for example, simply create a mobile-friendly version of your site and host it on your existing server. Then set up the .mobi version to redirect users to your mobile-optimized alternative.

This mobile version doesn't have to be fancy—just functional. Include some basic links and perhaps a feed that displays your most recent blog posts. Or better yet, talk to professional designer and have him/her create something that truly represents the value of your firm.

In addition, those working off of a WordPress platform can use one of their handy WordPress plugins to automatically *"read"* how the visitor is viewing your site and then redirect the user to your mobile or PC version, depending upon the user's platform.

The .mobi extension is still relatively new and there's some speculation as to whether it's going to catch on or die off. At this writing, there's simply not enough information to say for sure, especially when there are so many other ways to get your website mobile-friendly without having a .mobi domain. But one thing is for sure: Mobile is the next piece to this technological revolution. If you're not on board, you're missing out on clients.

11.7: *Press Releases*

There was a time when press releases were actually announcements for reporters and were never seen by the public. These releases were submitted to the media by companies hoping to get a little free publicity for their business in the way of a feature story in the news. Unfortunately, getting those efforts to actually pay off was a hit-and-miss proposition. If the announcement was considered to be really big news, it might get mentioned, sometimes weeks later and with no way to track the value it provided. If the announcement wasn't considered newsworthy, it stayed buried and your prospects never even knew the announcement existed. Needless to say, press releases weren't written very often.

But all that's changed.

Like everything else in our world, the Internet has given us a new set of rules to work with and press releases are now an integral part of a good online marketing campaign. Today's press release is picked up not just by the news media, but by the search engines as well. What's more, the media no longer has the power to stand between you and your target market—a fact they're well aware of—making them an ally rather than an obstacle to overcome. There are of course, still some strategies and guidelines to be followed when crafting your release. But, unlike the press releases of yesterday, these announcements will be well worth your efforts.

Writing Your First Press Release

A good press release still follows a certain format:

Contact Information: This includes the name and telephone number of the designated contact person as well as an email address and your website's URL.

Headline: Your headline should be bold, and enticing, offering a one-line summary of your announcement. *The Law Firm of Smith and Davis Celebrates 30th Anniversary in New York City with Free Open House* for example, or *Estate Planning Law Firm Smith and Davis Offers New Wealth-Building Seminar*. Notice that these two announcements aren't necessarily *"big"* in terms of media-worthy news. But, they are big news as far as your firm goes. That's what the new generation of press releases is all about.

Opening Paragraph: This paragraph includes your location, the date of your release and a quick statement—two to three sentences—that expands on your headline.

Body Text: Includes details about your announcement or event as well as reasons the public should take note. This is also where you'll include quotes, either from existing clients or employees of your firm.

Sub Headlines, Additional Body Text: In general, your press release should be no more than one page but if it needs to go longer, break up the text with sub-headings.

Summary: Wrap up your announcement and then restate your contact information with a sentence such as *"for more information, contact...."*

Company History: In a separate paragraph, write a short bio about your firm and its services.

So, how does this look as a finished product? Here's the text version of the press release we wrote to promote this book:

Best Legal Practices Announces New Marketing Handbook for Law Firms

San Diego, CA—Attorneys wanting to join the online marketing revolution need look no further. Best Legal Practices has just released the ultimate online marketing handbook and it's geared specifically for small law practices.

Because of the unique nature of the legal profession, attorneys have never had to be concerned with marketing in the past. But of course, with the social revolution we're now experiencing, it's become extremely important for attorneys to master the art of online marketing. This new book, *Dominate Your Market: The Attorney's Complete Guide to Online Marketing and Social Media*, provides a comprehensive resource on this topic. Complete with tools, tips and plenty of how-to's, *Dominate Your Market* explains online marketing in a step-by-step process and shows attorneys how to customize the entire process to suit their unique needs.

"We're so excited to release this book," said Robert Armstrong, co-author and co-founder of Best Legal Practices and the exclusive American Academy of Estate Planning Attorneys. *"We've spent years teaching estate planning attorneys how to modernize their practices and use marketing techniques to build a solid and profitable firm. This book drastically expands on that mission by showing attorneys of all areas of practice how to create and manage their own marketing campaigns in a step-by-step format. It really is the must-have book for small law firms."*

Dominate Your Market: The Attorney's Complete Guide to Online Marketing and Social Media will be available on Amazon in June of 2011. For more information about Best Legal Practices, contact Chelsea Wilson at **800-975-6448** or send an email to support@bestlegalpractices.com

About Best Legal Practices
Best Legal Practices is a professional company created to provide marketing and practice management solutions to the next generation law firm. The company is the vision of American Academy of Estate Planning Attorneys' founders, Robert Armstrong and Sanford M. Fisch. Armstrong and Fisch are also the co-authors of *The E-Myth Attorney: Why Most Legal Practices Don't Work And What To Do About It*. Best Legal Practices offers educational courses and SEO and SEM domination marketing services to its members. For more information, visit **bestlegalpractices.com.**

Press Contact:
Chelsea Wilson
Best Legal Practices
800-975-6448
support@bestlegalpractices.com
bestlegalpractices.com

But we're not done just yet. Today's press release has quite a bit more to offer than just text.

Making Your Press Release Social

In addition to being easier to publish, online press releases offer several advantages over their traditional, hard-copy counterparts. You can include links, for example, to your website, your landing page, a free report or anything else you deem relevant.

Online press releases can also accommodate: images, videos, PDF downloads and highlighted quotes, allowing you to create interactive and informative content on high-ranking press release websites.

In this release from the City of Long Island, you can see the traditional press release layout: bolded headline, copy introduction, body, etc. But, you also see a number of other features.

The city included a video, two graph images, and multiple downloadable reports in PDF format. They've also drawn attention to their core issue by highlighting an important quote in a sidebar.

This type of release is both eye-catching and intriguing. It offers the reader plenty of additional material to help support the announcement itself. If this were a private business, you might also see multiple links throughout the content, inviting you to visit the company's website, join their mailing list or some other relationship-building activity.

Also note that there are social media buttons included. So, the reader can instantly share this news by emailing it, printing or posting it to one of his/her own social media profiles.

Avoiding the Marketing Hype

Read a few press releases on the web and you'll quickly see a common denominator among those announcements. They all contain various buzz words and marketing hype. Words like *"cutting-edge"* and *"state-of-the-art"* are basically just filler words that marketers use because they sound better than *"good"* or *"great."*

Using some of these words sparingly is okay. After all, the whole point of your press release is to promote your firm, so telling your audience that you can offer a *"unique solution"* or that your firm is one of your city's *"leading providers"* of legal services is acceptable.

Just don't get carried away. Your press release should be informative, meaning that it contains something of substance. Filling the page with buzz words or legalese isn't really telling your audience anything. It's just a way to toot your own horn and you can bet that your readers will see right through that tactic.

Instead, tell them *why* you're one of the leading providers of legal services. Explain *how* that unique solution will solve their problem. Write clearly and honestly, so that your prospects know exactly why you wrote your press release in the first place. **Important Note: As with all messages to the public, be sure to consult your State Bar ethics rules before issuing press releases.**

What's Considered Newsworthy?

In the old way of doing things, press releases were only issued when something really big happened. Today's press release works much differently. You don't have to wait until you've won a big case or brought in a new partner to promote your firm. Quite the contrary, you can issue press releases for just about anything.

Like what, for example?

Well, if you're hosting an on-site seminar, that's worth a press release. If you've come up with a new way to bundle your services, promote that *"product"* to the public. Has there been a big change in the law? Hired a new associate? Published a new educational report? Attended special legal training? Received an award? All of these events are worthy of a press release.

The idea is to generate interest about your firm. Since you no longer have to rely on the media's *"acceptance"* of your news, you don't need to be as picky about what you promote. This reminds us of the movie *Arthur*, when Dudley Moore announces he's going to take a bath and his long-suffering butler quickly responds, *"I'll alert the media."* Clearly, you want your releases to have a little more value than this. But, you do still have a large amount of flexibility in what you choose to announce.

So, look around! What's going on in your firm? What new services have you added? What features do you offer on your website? Think about it and then write a press release to promote it.

Distributing Your Press Release

So, now that you know how to craft an effective press release, the next question is where do you submit it?

Once again, the new rules of press releases come into play. You don't have to worry about digging up contact information for individual reporters. Instead, there are a variety of press release services right on the Internet. Get your release listed on some of these and the rest, as they say, is cake.

PRWeb—Probably one of the most notable press release distribution sites, PRWeb offers several distribution packages to match your needs. All packages ensure placement on the major news sites such as Yahoo!News, and guarantee indexing by the major search engines. Higher-priced packages include: video placement, social bookmarks, attachments and the other great extras we mentioned above. To register, go to prweb.com.

Online PR Media—Formerly Online PR News, OPM has both free and paid plans for your convenience. The free account offers you the ability to issue plain-vanilla releases with one live URL in the media contact area. You can also display your website in an iFrame. The paid versions are where you'll get the bells and whistles. For as little as $49 per release, you can get a headline tweet, add up to three file attachments, three anchor text links, and a PDF version of your release published on DocStock. To register and start submitting, go to onlineprnews.com.

PR.com—This service offers three levels of membership: The silver level is free and allows you to distribute both paid and free releases as well as post job openings in the **PR.com** job search website. The gold level is $199 per year and includes a full company profile, a listing in the **PR.com** business directory, free and paid press release distribution and four free anchor text links per year. The platinum level is $499 per year and includes all the features of the gold level as well as unlimited product/service postings and ten free anchor text links per year.

PRNewswire—PRNewswire is considered to be a premier online distribution service. Pricing starts at $195 per year with additional charges per release. Features include: audio and video webcasting; editorial calendar service and extensive customer support; and guidance on your press release campaigns. Go to **PRNewswire.com** to register.

MarketWire—MarketWire has a unique business model in that it allows you to target your press releases geographically. That means if you want your releases to focus on news services in California or even just San Diego, for example, you can do just that. Select your option and then, in addition to national publications such as The New York Times and the AP and Dow Jones newswires, your release will be sent to a variety of online and offline publications in your chosen area. Call for pricing or go to **marketwire.com** for more information.

Chapter 12:
Using Social Media to Build Your Community

You can't have an online marketing campaign without running into social media and for good reason: Social media has literally changed the way we interact and communicate on the web.

It's also changed the way we market and given us a direct line to those targeted clients we've been trying to reach.

Now, if you're thinking that social media refers to third-party applications such as Facebook and Twitter, you'd be right. But that's not all that social media is… not by a longshot.

12.1: What Is Social Media?

There are actually hundreds upon hundreds of different social media sites, all designed to target a specific interest, topic or group of people. Each SM site also serves a specific purpose, whether it's to educate, organize or simply give users a way to communicate and share information.

In fact, that's probably the best definition for social media:

> *A collection of technologies designed to encourage participation,*
> *communication and collaboration within a given community*

When we say *"encourage,"* we're being conservative. Social media entices its users. It invites them in and immediately gives them a sense of belonging and of purpose. It almost *demands* participation. If you hang around at a given SM site for any length of time, we're betting you'll find at least one thing that piques your interest, catches your eye, or inspires a comment.

And that's the magic of social media. It makes all the normal social barriers like social status and location irrelevant, and allows people to come together based on a mutual need or interest.

Unfortunately, there are still a number of business owners that see social media as simply another way for employees to kill time at work when the boss isn't looking. But this attitude is keeping these employers from tapping into one of the most beneficial marketing strategies of our time. Social media isn't just about sharing pictures of your kid's soccer game with family members. Quite the contrary! There are a number of lawyerly things you can do with social media and certainly quite a few marketing tasks you can perform. This is why we're spending time on this topic now.

So, let's start by looking at the different forms of social media technology:

12.2: Different Types of Social Media

Social media comes in a variety of shapes and sizes, including: forums, blogs and micro-blogs, networking sites and wikis plus photo-sharing and bookmarking sites to name just a few. In a 2010 *Business Horizons* article,

Andreas Kaplan and Michael Haenlein attempted to classify social media sites by dividing them into six distinct categories:

Blogs/Micro-Blogs
This category is focused primarily on communication. As such, it provides the user with several ways to share information and deliver content. Sites in this category would include personal or corporate blogs—often run on popular platforms such as Wordpress, Blogger, Xanga or Typepad and micro-blogs (Twitter, Foursquare and Tumblr).

Collaborative Projects
This includes wiki sites, such as Wikipedia, as well as document sharing sites such as: Dropbox and Google Docs. Blogs that are managed by multiple authors would also fall into this category as well as social bookmarking sites like: Digg , Delicious, StumbleUpon and Reddit (discussed in depth in Chapter 12). The point of these sites is to allow users to participate in the creation of content, whether it's actual articles such as what you'd find in a wiki or blog, or simply sharing links to content and rating its value (bookmarking sites).

Social Networking
This category includes sites that encourage community-building by creating personal profiles, connecting with your own *"network"* of friends and family and sharing information. Sites in this category would include Facebook, MySpace, High5, and LinkedIn.

Content Communities
A content community is defined as a place where users can share media content with each other. Examples of sites in this category would include the photo-sharing site Flickr, YouTube (videos), Scribd (PowerPoint presentations) and Podbean (podcasts). Opinion sites, such as Epinions.com, Q&A such as eHow and Yahoo Answers, article directories and content sharing sites such as: Squidoo, HubPages, and Associated Content would also fall into this category.

Virtual Social Worlds
This category represents sites that have literally built alternate worlds where users can shop, eat, and interact. Perhaps the most popular version of this kind of site is SecondLife, a virtual world complete with a monetary system that can be converted into US dollars on the stock exchange.

Virtual Game Worlds
Like their social brethren, virtual game worlds include the multiplayer online role-playing games (MMORPG) such as World of Warcraft and EverQuest. The primary difference between the gaming world and the social world is that players are held to strict guidelines within the gaming environment to support the roles and storyline of the game itself.

Now, for purposes of this book, we're going to focus on the first four categories and leave the virtual worlds for another time. Just so you know, these virtual worlds can offer big benefits with the right marketing campaign. Many large corporations have an active presence in social worlds such as SecondLife, where they interact with potential customers, promote their brand, and even sell virtual products and services.

Likewise, the gaming world isn't necessarily just for fun either. Pizza Hut, for example, allows players in the EverQuest II virtual world to order a real pizza for real delivery from its in-world establishment.

How's that for connecting with your audience?

Back to the topic at hand. In addition to these basic categories, you can also add aggregators which are designed to pull your multiple profiles into yet another profile so your *"followers"* can see where you are and what you're doing, all in one handy place. There's also an endless supply of apps and plugins that can customize your social network profiles by adding an event calendar, for example, or a virtual fundraising center where you can support your favorite cause and help raise money in the process.

You can track your followers, the topics they're discussing, and even watch to see how your firm or brand is trending across the social networks, all with just a few clicks—if, of course, you have the right application.

As you can see, social media is more than just a chat room for teens and college students. Clearly, it's here to stay. The question is: Are you ready to embrace Web 2.0 and make the most of the rapidly changing technology to dominate your market for legal services?

If your answer is yes, then read on. We're going to show you how.

12.3: Getting Started With Social Media

The first step in *"going social"* is to set up your profiles. Each SM site is a little different but most will require you to choose a user name (or your email address) and a password.

For simplicity, use the same password for all your SM accounts and choose something that you can easily remember but that's difficult for others to guess. You might also consider swapping numbers for letters, such as a 1 for an L or a 0 for an O. For example, instead of the password *"envelope,"* you could use *"enve10pe."*

The second thing you'll need to do is choose a profile image. With the exception of your personal Facebook profile, you should use your firm's logo for this purpose. As an alternative, you could use a candid photo of you in the office or behind your desk.

The point of this image is to support your firm's brand. So be sure the photo you choose matches the image you want to project. One final tip on this subject: if the photo is of you, make sure you're smiling.

We realize that many attorneys prefer the more somber look in their professional photos but remember, it's not other attorneys you're trying to connect with. It's your ideal client. Chances are, they'd prefer to consult with someone who doesn't look quite so intimidating all the time.

Now, having said that, here's a quick overview of the SM sites you'll want to target first:

Facebook

Facebook is fast becoming THE place to be in the way of social networks. It currently boasts more than five hundred million active users. Each one of those five hundred million users creates an average of ninety pieces of content each month and shares that content with an average of one hundred and thirty friends. In total, Facebook's users spend over seven hundred billion minutes on Facebook every month. According to the analytics company, Hitwise, Facebook became the most popular website in the United States in 2010, surpassing even Google.

What's more, Facebook's fan pages and groups make it possible to target your ideal client fairly easily, giving us yet another reason to add Facebook to your digital marketing campaign.

To do this effectively, you need to understand how Facebook works and where your efforts should be focused.

Your Facebook profile will actually have two components. The first is your personal profile and the second is your firm's fan page. Let's start with the personal profile.

Setting Up Your Personal Facebook Profile

To get a fan page and enjoy all its features, you need to first set up a personal profile.

This profile allows you to *"befriend"* other users, and then those friends can view your Wall, see your status updates and pretty much everything else that's on your profile. There is a way to limit the amount of information that each *"friend"* can see, but it's a tedious and counter-productive process. Your time is better spent doing other things.

To set up your profile, you'll want to create a new Facebook account (if you don't already have one) and then do the following:

- Upload a profile image—this should be a picture of you, something candid and friendly... not that stodgy old picture you use for the lawyer directories and newsletters.

- Complete your contact information, including your address, phone number, fax, website URL, email address and, of course, a short blurb for your bio.

Be sure to set up a personal account—not a business account—or you won't have access to all the features that Facebook has to offer.

Now, what happens if you already have an existing Facebook account that you use for family and friends? Should you tie your firm to your personal account? Read on for thoughts on both sides of this argument.

Separate Identities: Should You Maintain Multiple Facebook Accounts?

There is a growing and on-going debate about the need for separate Facebook accounts. Those in favor argue that it's the only true way to keep your personal life separate from your professional persona. And they've got a point.

After all, no one wants to bombard their family members with daily posts relating to business. You certainly don't want your potential customers seeing the family photos taken at Aunt Jean's birthday party or the video of your daughter's first violin recital. Those are, well... personal. It makes sense to keep the two separate.

To counter this argument, Facebook came up with Fan Pages, which allow you to create a separate Facebook personality for your business. Fan pages are linked to and maintained through your personal profile, but other users can't see that link and therefore, can't make the connection.

Problem solved, right?

In the past, users couldn't *"like"* another page from their fan page—that had to come from a personal profile, making it difficult to keep your personal account and your business life separate. Facebook recently changed that policy and as of this writing, you can now become a fan of another page without exposing your personal profile.

Why does that matter?

When you *"like"* a page, those connections usually result in friend requests, which means that you'll have to give those target clients access to your personal profile if you want to accept the request.

And suddenly, we're back to mixing your personal life with your professional side, something that many users just don't want to do.

So, why all the fuss? Why not just create two separate accounts?

Many users do just that. But, you should know that having multiple accounts is a violation of Facebook's Terms of Service. Thus, the on-going debate. So, while we like the idea of keeping things separate, we can't advise you to create that second account.

You'll have to decide what works best for you.

Building Your Firm's Fan Page

Your fan page is a powerful marketing tool and one of the main reasons you're joining Facebook in the first place. So, pay attention to this section carefully. It can have a big impact on your overall online success.

To create a new page, log into your Facebook profile and click *"Home"* in the upper right-hand corner. Scroll down to the *"Ads and Pages"* option in the left-hand navigation menu and click. If you don't see *"Ads and Pages,"* choose *"More"* to reveal other options.

Hint: If you still don't see the page option, go to www.facebook.com/pages.

Click the *"Create Page"* button in the upper right-hand corner:

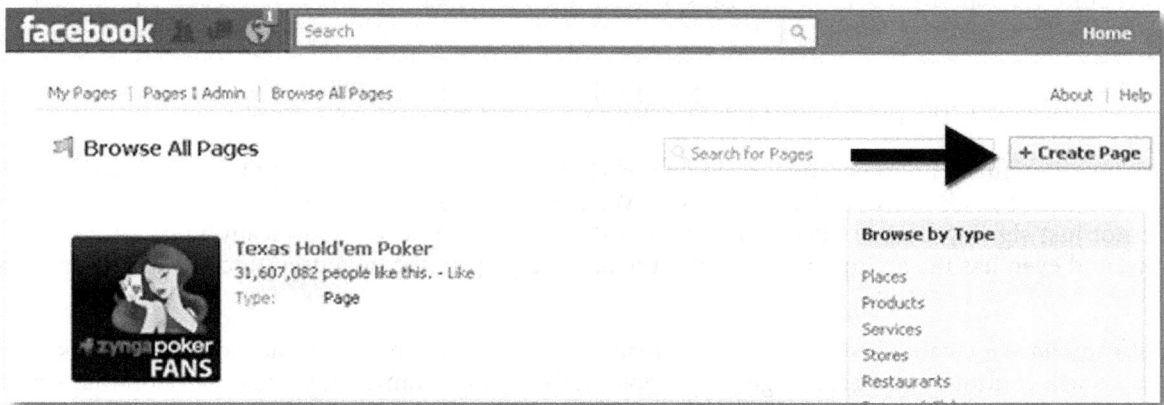

and then choose *"Local Business"* and a page name. Generally, the page name should be the same name as your firm.

Check the box that says *"I'm the official representative."* Then, click *"Create Official Page."* Viola! You've just created your first Facebook fan page.

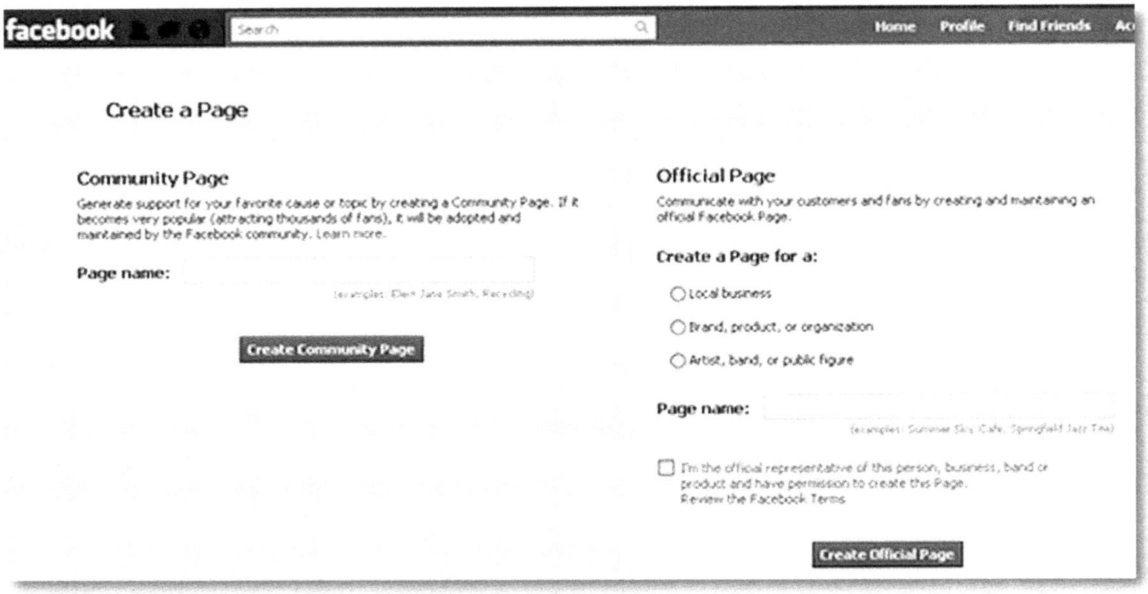

Now comes the fun part. You need to customize it. In addition to uploading a separate profile image and plugging in basic contact information about your firm, the fan page offers you a unique marketing opportunity that you can't get with your personal profile.

Until very recently, fan pages could be customized using Facebook Markup Language (or FBML for short). This was Facebook's version of HTML. For the most part, it was a fairly easy way to create a customized tab on your fan page.

Unfortunately, FBML couldn't handle dynamic content, something that many developers saw as a big drawback and Facebook listened. The result is that FBML is being replaced by the iFrames application.

Now, iFrames is actually not new. The well-known HTML tag essentially allows you to create a web page at one location and insert it somewhere else on the web. What this means is that your new customized fan page can contain not just sign up boxes, videos, buttons and downloads, but also dynamically-updated content, Flash elements, and even has the ability to personalize the landing page based on a user's location or other track-able value.

If you already have a customized landing page created in FBML, not to worry. Facebook assures its users that those pages will continue to work. But given the potential of using iFrames, it stands to reason that you'll want to incorporate this new feature into your fan page presence.

To start using iFrames, you'll need to first become a verified Facebook apps developer.

This is easier than it sounds.

Go to facebook.com/developers. If you haven't already verified your Facebook account, you'll see the following screen:

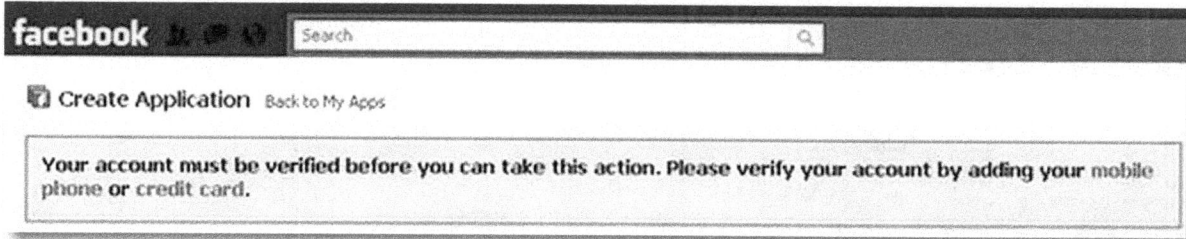

Verify your account as instructed and then come back to this page. You'll see a button that says "*+ Set Up New App.*" Click it.

Give your app a name, agree to Facebook's Terms of Service and click "*Create App.*"

After passing a security screen, you'll see the app creation screen, starting on the "*About*" tab. This is where you enter basic information about your app, i.e., the name (if you want to change it), a brief description, contact email address for your users, etc. You can also insert a logo here for your app—the smaller image is what will show up next to the link for this page in the navigation menu on your fan page. This is similar to a favicon (at least 16 x 16 pixels) so use an image of your logo or something comparable.

Don't worry about the larger image. This is what would show up if you ever decided to submit your app to the app directory. But, since your app is not something others will want to use, this image won't be needed.

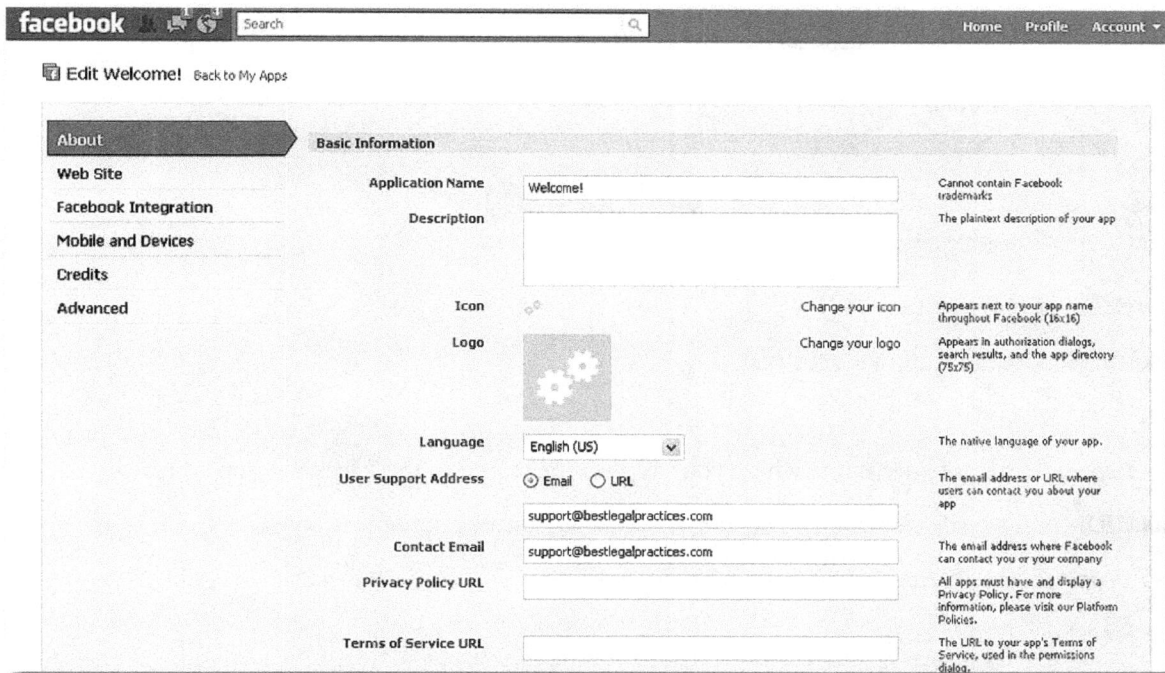

Now click the "*Facebook Integration*" tab.

This is where you tell Facebook how to find your app.

And here's how to fill it out:

Core Settings: Already completed for you.

Canvas:

Canvas Page—Create a folder name where your app will be stored in the Facebook apps directory.

Canvas URL—Use your own root directly here (i.e., we use bestlegalpractices.com).

Canvas Type: Choose *"iFrame"*

iFrame Size: Select *"Auto-resize"*

Page Tabs:

Tab Name: What do you want to call your tab in the navigation menu?

Page Type: Choose *"Frame."*

Tab URL: The actual page where your coding is stored on your domain.

Now, if you're a little confused, let us explain:

To use iFrames, you need to create a page and host it on your own domain, just as you would if you were creating that web page to act as part of your website. The iFrames tag then pulls the coding from that page and inserts it into your fan page.

That's why the *"Tab URL"* contains a file name from your firm's domain. This is where you've stored the file you've created to act as your fan page.

Obviously, you'll want to have this file created before you set up your app on Facebook. For this, you're going to need someone who is familiar with web design and coding.

Depending upon the features you want your fan page to have, your designer might need to know HTML, CSS, Java, Flash and a whole host of other scripting and programming languages to make your fan page work.

Certainly, if you have someone in your office that can write basic HTML and give you a simple fan page, then by all means go for it. But if you don't have such a staff member or you want something a little more dynamic, you'll need to hire a professional web developer to handle this piece for you.

Well, let's say that you'd like to create a page that shows a nifty little video about your firm or offers a free report to those who sign up for your mailing list and let's say that you want this page to be the first thing new visitors see—with iFrames, you can do just that. Create the page the way you want it and then set it as your default page for new visitors.

You can also create a custom profile image that includes not only your logo or personal photo but also some branding concepts below. Facebook allows your fan page profile image to be 200 x 600 pixels. That means you can put your logo on top and then include a catchy tagline or something equally flattering below it. Facebook won't resize the photo as long as it stays within these dimensions. To see what we're talking about, take a look at our profile image on our fan page. Go to on.fb.me/blpbook.

Other options for your fan page include: pulling your blog's feed onto your Wall, connecting your fan page with your Twitter account and perhaps setting up a separate page that lists all your upcoming seminars and events. The goal is to make your fan page intriguing and inviting to new visitors so look for ways to add some interaction, give it some color, and make it stand out. This is one of the primary tools we'll be using to promote your firm.

BestLegalPractices

Winning Strategies to Dominate Your Legal Market

Law school taught you how to be an attorney! We will teach you how to run your business...

twitter.com/bestlglpractice

BestLegalPractices.com

Suggest Us to Your Friends

Facebook Groups

Facebook has had the *"groups"* feature around for quite some time. But, it's only recently that this feature was revamped to become a more integral part of your Facebook experience.

The idea behind the new groups format is that it is a better representation of how you segment your life (better than your friends list, to be exact) and it gives Facebook more accurate data with which to measure the performance of its services.

The difference between the fan page and a group is somewhat blurred. The fan page is meant to represent a separate entity, be it an organization, a corporate brand or—as in this case—your law firm. Groups on the other hand, are intended to represent a collective: a group of like-minded people coming together for a mutual reason or cause.

But that doesn't mean that there aren't fan pages out there representing causes and interests. Nor does it mean that companies and law firms don't have groups. Both are true.

This new revamp may have been intended to define the lines between pages and groups a little better. But, there's nothing to say that you can't form (or join) a group to support new tax legislation, act as a platform for consumer fraud, or simply provide a community for those who are going through a divorce.

See how this works?

If you consider your fan page to be your main Facebook attraction, then groups is one of the best ways to find your audience.

So, that's where we're going to start. Log into your Facebook account. Look at the left-hand navigation menu from your *"Home"* page. Click on the link that says, *"Groups."*

From here, you can search for groups that relate to your topic of interest. A word of advice: get creative with your search strings.

The term *"divorce,"* for example, returns a multitude of possible group matches. But, you could refine that search with terms such as *"divorce recovery," "pet custody"* and *"children of divorce."* All these terms return results that you might not have seen otherwise. They give you a great opportunity to focus your marketing on specific aspects of your practice area.

Should you not find the group you want or still feel that you'd like to create your own, you can do so. Just click the *"Create Group"* button at the top of the Groups page.

Groups give you several features that you don't get with fan pages. For example, you can send a mass mailing to your group and anyone who has joined that group gets the email. This is a great way to promote seminars or webinars you might be doing on a related topic. Just be sure to use this feature wisely so that your members don't see your alerts as *"spam."*

Also, while you're free to use this platform to promote your website, free reports and services, be sure to balance that self-promotion with plenty of non-promotional material. Remember, you're here to build relationships. This is first step in creating a lasting and loyal client base.

You can also share documents with your group, engage in a group chat, and receive alerts anytime someone else joins your group or posts something new for the group to see.

The great thing about groups is that it's completely voluntary, so anyone joining actually *wants to be there*. Thus, this is a good place to start finding your ideal clients.

You can join (or create) as many groups as you like. So get busy. This is free advertising at its best.

How To Increase Friends/Fans/Followers on Facebook

The whole point of having a presence on Facebook is, as we've said many times, to build relationships. So, the more friends and fans you have, the more relationships you have an opportunity to build.

There are several ways to attract new friends and fans. You can choose the ones you want to implement, depending upon how you've decided to use your personal profile.

- **Tip #1**—Use the Facebook search bar to search for keywords and phrases related to your practice area. The results will include people, groups and pages, so you can browse through and see which ones might fit with your objectives. Just remember that if you choose to *"friend"* your potential clients, they'll have access to your personal profile. If this isn't a problem for you, then feel free. But, otherwise, you might want to limit outside access to your fan page and group memberships.

- **Tip #2**—Join related groups and create a few of your own. Groups are a great way to meet other Facebook users that have an interest in your area of expertise.

- **Tip #3**—Promote your fan page on your blog or website as well as in all of your firm's correspondence and marketing material. There are some great widgets out there to highlight your fan page on your website. (We discuss these in the website design chapter, in case you're wondering.) If you want to shorten the link for printed material, try a URL shortening service such as bit.ly.

- **Tip #4**—Create a fun and inviting landing page for your visitors. Include a compelling video and make sure your page has plenty for visitors to do and see. **Hint:** Remind your visitors to *"Like"* your page.

- **Tip #5**—Take advantage of apps. There are thousands of applications that have been developed for Facebook. Many of them can really add to the visibility of your fan page. The Threadless app, for example, allows users to comment on items displayed on your fan page. These comments are then incorporated into the user's Wall, instantly creating a viral effect for your services. Now, if you're thinking that this would only work with a retail type of business, think again. Your free reports would work and give you an opportunity to test the value of your giveaways.

- **Tip #6**—Invest in Facebook's PPC advertising program. Facebook's pay-per-click program is quickly surpassing Google Adwords in many respects, simply because of Facebook's ability to let advertisers drill-down their target markets. Think about all the personal questions that Facebook allows users to answer: age, interests, marital status, income range… All this data can be used in your PPC campaign to really target your ideal client. For more on Facebook's PPC program, go to the Pay-Per-Click chapter in Part 3 of this book.

Twitter

There was a time when Twitter was seen as merely another entertainment application—a hybrid of the old Internet chat rooms and the more modern instant messaging platforms.

But today, Twitter has evolved into a powerful marketing tool. Like other social media applications, Twitter can be used to boost your firm's bottom line.

Setting Up Your Twitter Profile

In its original layout, Twitter offered a wonderful way to incorporate your firm's brand with your Twitter account. Unfortunately, they've since resized the user interface and the result is that all those custom backgrounds no longer fit the way they should.

But that doesn't mean you can't still customize your Twitter background. You just have to be a little more creative about doing it.

The first thing you'll want to do is create a new Twitter account if you don't already have one. Try to use a form of your firm's name as the user ID. Twitter limits you to fifteen characters for this so you may have to abbreviate a bit to make it work.

Once you've signed up, go to go *"Settings"* in the top, right-hand menu. Click the *"Design"* menu option and scroll down just a bit where you'll find a button that says *"Change background image."*

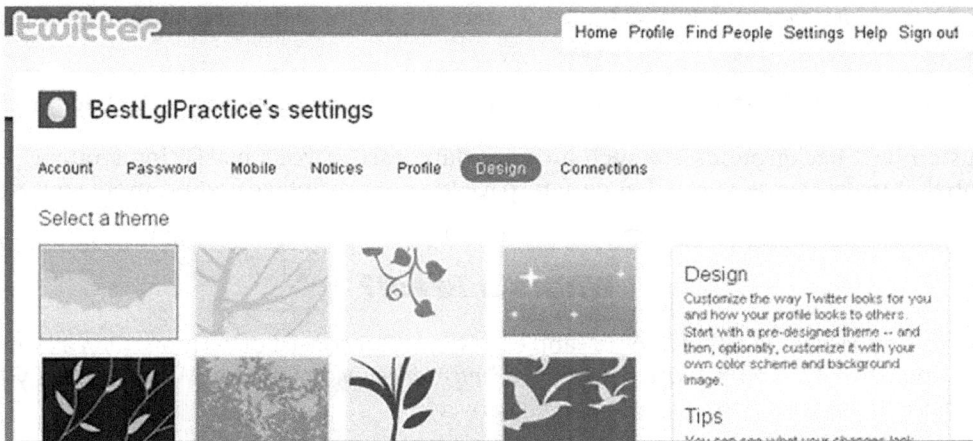

To create a custom Twitter background, your file needs to be smaller than 800k in size and must be in GIF, JPG or PNG format.

The new Twitter uses a fluid layout, meaning that the size of the browser window (and thus, the screen resolution) can affect the overall appearance. In addition, there is a toolbar that now sits on top of your page that's about 40 pixels tall.

The old Twitter had open space of 250 pixels on each side of the user interface. This was considered prime design space and could be used to create a very *"branded"* background. That space has been greatly reduced, to just over 110 pixels on each side.

Also, the right-hand sidebar is now semi-opaque, meaning that with the right color scheme, you can have some additional design elements show through that sidebar.

To create a background, you'll either need to enlist the help of a professional designer or, if you want to do your own, you'll need to invest in a graphics program such as Photoshop and start with a background that's 1920 x 1200 pixels in size to accommodate viewers with larger screens. That said however, most users will be viewing at either 1024 x 768 or 1290 x 800, so you'll want to test your background at those resolutions to make sure everything is copacetic.

To maximize the limited design space, we recommend using the left side for your firm's name, turned sideways so that you can use bigger fonts, but fitting it within an area that's about 45 x 500 pixels. The right-hand side can accommodate a large version of your logo or perhaps a relevant graphic, but you'll need to set the design colors so that this image shows through the semi-opaque sidebar.

For examples of this, check out these Twitter accounts:

twitter.com/bestlglpractice

twitter.com/mikosoft

twitter.com/copyblogger

So, now the question becomes, why would you want to customize your Twitter background?

Keep in mind that as of April, 2010, Twitter had over a hundred and five million users with new users signing up at the rate of *about 300,000 per day*. That's a lot of users and they're averaging about fifty-five million tweets each day.

Now, imagine if your firm had just a piece of that activity. What do you suppose it would do for your business?

This is why customized backgrounds are such an important part of your marketing strategy. With so much visibility potential, it makes sense to use that design space to your advantage and promote your firm.

What's a Tweet?

Each entry you post on Twitter is called a *"tweet"* and each tweet can consist of no more than a hundred and forty characters. This, incidentally, is why Twitter is called a *"micro-blogging"* platform. Whatever it is you want to say, you must be concise to get your point across in a single tweet.

Because of the small character limit, users have learned to maximize this space by integrating abbreviations, emoticons and other forms of shorthand into their posts.

So, the message

> *We've just posted a new article on our blog! Find out the five things you should know before filing for a divorce and protect your assets!* **www.yourfirm.com/page.html**

… becomes…

> *New article on our blog! Find out the 5 things u should know b/f filing 4 a divorce & protect ur assets! bit.ly/blog1*

…and looks like this:

This version actually still has seventeen characters to spare. So you could modify it a bit more to include additional information if you wanted to. Once you're happy with your tweet, just click the *"Tweet"* button.

The odd reference link you see is courtesy of bit.ly, a URL shortening service that's both free and allows you to track the traffic generated by your shortened link. You can also modify the last half of that link. So, in this example, we used *"blog1"* as the name of the page.

We also used **bit.ly** to shorten the link to our Facebook fan page for this book, so the link we provided earlier—on.fb.me/blpbook—represents only hits to our fan page that were referred from this book. Not only does this save space and serve as a nice way to see how many people make it to a specific link from a specific source, but it also avoids asking you to type the real, and often unwieldy links, such as the one to our fan page—**facebook.com/bestlegalpractices.**

There are actually a number of different URL shortening services and given the restricted character limit in Twitter, it makes sense to use one from time to time. We obviously like bit.ly, but we've included a list of others you can check out in the appendices section of this book.

To tweet with another user, you simply have to include their user ID within your tweet. So for example, if your ID (the name you chose when you set up your account) is BillJones then, your Twitter ID is @BillJones. If someone wants to send a tweet to you, they just add *"@BillJones"* anywhere in the tweet. Likewise, you can respond back or start new conversations by including the Twitter ID in your post.

Your Twitter wall will display your tweets as well as tweets from those you follow (whether you're mentioned or not) in reverse-chronological order. If you want to see only your tweets, click *"Profile"* from the menu across the top of the screen.

If you want to see only tweets that mention you, click the *"@Mentions"* link in the navigation menu just below the *"what's happening?"* text box on your wall.

What Should I Tweet About?

The great thing about Twitter is that it gives you an opportunity to combine a little self-promotion with casual conversation, allowing you to create a relationship with potential clients that doesn't strictly revolve around what you want them to buy next.

Yes, we said *relationship*. Social media is an ideal tool for promoting your firm and building your business but you have to be *"social"* for it to work. Simply making post after post about the services you have for sale will get you nowhere and this is especially true in the Twitterverse. If you're not willing to get involved in conversations and *"chat"* with your followers now and then, then you should skip this section altogether because you'll only be wasting your time.

Still here?

Okay, good. Let's go over some dos and don'ts for posting tweets:

■ Your Twitter *"wall"* shows you recent tweets from the users that you're following. It is perfectly acceptable—and in fact, encouraged—for you to respond to some of those tweets, even if it has nothing to do with your services and you've never met the person before. For example, if John Smith tweets about feeling overly stuffed after the holidays, you could simply respond to that tweet by agreeing—*@JohnSmith Ditto… way too much food!*—or *"retweet"* and add your own comment—*wonder how many miles I need 2 jog 2 counter pumpkin pie? RT @JohnSmith Still feeling stuffed from the holidays!* In both of these examples, you've *"mentioned"* John Smith's ID (@JohnSmith) so he'll see your reply and this is how Twitter conversations begin.

■ Retweeting is also a great way to show your support or appreciation for another's post, even if you don't add your own comments. If you don't have anything to add, just hover over the tweet with your mouse and you'll see a small menu pop up below the tweet. The options are *"favorite," "retweet"* and *"reply."* Click *"retweet"* and Twitter will take care of the rest.

- Anytime you post something new to your blog, add something to your website, create a new issue of your newsletter or anything else that promotes your firm, you should tweet about it. Type something short and to the point. But, be sure to make it intriguing. For example: *Learn about the new tax laws in this month's newsletter!* Or: *On Our Blog: When Ignorance of the Law is a Good Defense.* Of course, add your customized link.

- In addition to your own content, use Twitter to point to other worthy articles, videos, and other media that you find interesting or that relate to your practice area. A news article about the divorce rate dropping after ten years of marriage would make a great tweet for a divorce or family law attorney. Likewise, a study on the effects of genetics on criminal behavior would be appropriate for a criminal defense lawyer. You don't have to add anything to the topic or create an article of your own. Just tweet about it and provide a link.

- Now for the hard part: You're going to have to add a little personal touch to your tweets. That means, throwing out a few posts that simply show your human side. Talk about your run in the park, planning the office Christmas party, or the fact that tonight is spaghetti night with your in-laws. Remember, the point is to connect with your potential clients and the best way to do that is to let them see the part of you that isn't Bill Jones, Esquire, Attorney and Counselor at Law.

Incidentally, the most *"retweeted"* words include **free, how to, blog post, check out** and **10**, suggesting that your readers are looking for links to good content on your website.

You should make it a practice to post several times a week—at least once a day if you can manage it and maybe a bit more in the beginning—because you're trying to build a following.

Now, speaking of that *"following,"* how do you get other Twitter users to follow what you're saying?

How To Build Your Following on Twitter

Your *"followers"* are the Twitter equivalent of Facebook's fans and friends. These are the people who have chosen to follow your posts, meaning that your tweets show up on their wall. Some users *"auto-follow"* meaning they'll follow you when you follow them, while others don't.

Ultimately, your goal is to attract followers who could end up being potential clients. But, in the beginning, you just need followers to help promote your firm and your tweets.

Fortunately, there are a number of ways to increase your following, some of which are extremely easy.

As we've already mentioned, the more people you follow, the more likely you'll stumble across those who follow back. So, the first thing you want to do is find some people you'd like to follow.

There are several ways to do this.

- **Tip #1**—Search a keyword or phrase in the search bar at the top of the Twitter page. A quick search, for example, returns a list of tweets by real estate agents, some investment experts and a few headhunters looking to fill open positions.

Using this example, it might be a good idea to follow the real estate agents in your area if this is your niche since they could likely refer some business your way. Ditto for the investment experts. But don't stop there. Click on the profiles of those you've chosen to follow and see who's following them. Remember, their followers are people who *chose* to have posts about real estate (or whatever your area of specialty might be) displayed on their wall. So, chances are, these followers would be interested in your posts as well. Follow them and use some of the tweeting ideas we discussed in the previous section to start up conversations and make connections.

- **Tip #2**—Use hashtags in your posts. Hashtags are keywords that relate to your post, marked with a #. This makes it easier to track what's trending on Twitter at any given time. So, a tweet about the death penalty could be marked with #deathpenalty to ensure that those interested in this particular topic can track the hashtag and see your post.

You can see which hashtags are trending by looking at the right-hand side of your Twitter homepage. About halfway down, you'll see a column called *"Trends."* This represents what people are tweeting about the most at that very moment.

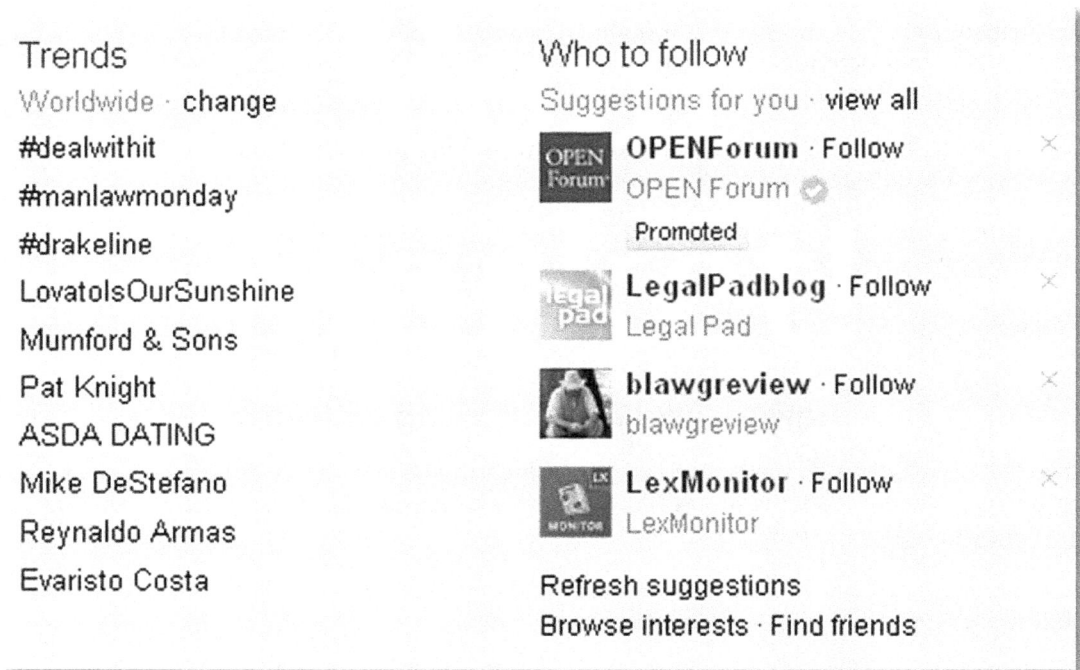

Incidentally, at this writing, the hashtag death penalty was tweeted eighty-eight times within the past twenty-four hours. Those eighty-eight tweets generated 80,938 impressions and reached an audience of 61,766 followers. How did we get those numbers? We'll show you this and some other great Twitter tools in the next section.

- **Tip #3**—On the right-hand side of the Trends column, you'll see another column called *"Who to Follow."* These are Twitter's recommendations for you, based on the people you currently follow and your hashtags. These suggestions will change periodically and new ones will be added as you follow those recommended on the list.

- **Tip #4**—Tweet. Don't ask us why, but the more you tweet, the more people tend to follow. Maybe they're searching for the topics you're tweeting about or maybe your tweet pops up because they're following someone who's following you. Whatever the reason, activity breeds followers, and it works every time.

- **Tip #5**—Promote your Twitter activity on your blog or website. There are a couple of different ways to promote Twitter on your website. We discuss all of them in the section on website design. If you haven't already implemented those changes, now's a good time to do that. Don't worry! We'll still be here when you get back.

- **Tip #6**—Add your Twitter ID to all of your firm's corporate identity materials, such as: your business cards, your letterhead, Facebook fan page, brochures, newsletters, and, of course, the signature line in your emails.

- **Tip #7**—Offer to write a guest post or article for a popular website or blog. Find other Twitter users that have a good following (and a relevant website or blog) and then offer to write a guest article for their site. They'll enjoy the exclusive insight provided by an attorney and you'll benefit from their promotional efforts on Twitter.

Great Twitter Tools

Once people (marketers, in particular) realized how much Twitter had to offer, a whole host of new apps and services popped up to help users maximize their Twitter usage and monitor performance. Initially, you may not need all these tools and this is by no means a complete list. There are literally hundreds of third-party applications designed to optimize Twitter usage. Be aware: Twitter apps come and go just like everything else on the Web. But this list will certainly give you a starting point and let you see just what you can do with all those tweets.

- **HashTracking.com**—We mentioned the importance of hashtags earlier and here's a good reason to use them. **HashTracking.com** allows you to search a particular hashtag and see who's using it and how popular the topic is. It also gives you a global overview, i.e., number of tweets, impressions, and viewers within a twenty-four-hour period.

- **TwitScoop.com**—TwitScoop can also track hashtags. It allows you to choose different time frames: most recent, past week or past month. You can view the *"Hot Trends"* to see what's happening now or you can search your own hashtags if you don't want to wade through all the currently trending topics.

- **TweetBeep.com**—TweetBeep is the Twitter version of Google Alerts. You tell the service what keywords you'd like to receive alerts for and TweetBeep will email you when those keywords show up in others' tweets. This is a great way to see who's talking about you, your website, or just your general topics of interest. You can select up to five keywords for free or use one of their paid packages to receive more alerts.

- **TweetDeck.com**—TweetDeck allows you to monitor both your Facebook profile and all your Twitter accounts (should you have more than one). You can create lists, manage your tweets, and monitor your mentions from one single screen.

- **TweetOClock.com**—TweetOClock is a unique little tool. It is ideal for someone who wants to do some target tweeting. If you've taken the time to research power Twitterers and want to make contact, TweetOClock can tell you the best time of the day to contact that user. Simply type in their Twitter ID and the service will show you when they're most active on Twitter.

- **TwitDoc.com**—One drawback of Twitter is that you can't include attachments—only links. TwitDoc is a desktop client that solves this issue by allowing you to upload media (images, documents, video, etc.) not already posted somewhere else on the web and tweet about it in one single step. TwitDoc also automatically shortens the URL for you so you don't have to switch over to a shortening service to complete your tweet.

- **JustTweetIt.com**—This is a Twitter directory of sorts that allows you to distinguish yourself by topics so that other like-minded Twitterers can find you. It also works in reverse. You can use it to find Twitter users that might be interested in what you have to say.

- **Twitter Friends Network Browser** *(www.neuroproductions.be/twitter_friends_network_browser/)*—This is an interesting service that allows you to view your followers as well as your followers' followers, and so on, and so on, and so on. Use this app to find new people to follow based on similar interests and activity levels.

LinkedIn

With so much buzz about Facebook and Twitter, you might be wondering if there's really any point in working with other social networks, such as LinkedIn. The answer is a definitive YES!

LinkedIn was created specifically with businesses in mind and can serve as a powerful networking tool for you and your firm. In fact, LinkedIn is considered to be the largest *professional* social network currently available and boasts a whopping eighty million + users.

Now, the great thing about LinkedIn is that most of the users are professionals as well—middle and senior management level individuals who have an average household income of $100,000 or more per year.

In short, this is likely a demographic you'll really want to target. So, how do you make the most of your presence on LinkedIn?

Getting Started With LinkedIn

The first thing you'll want to do is complete your profile as much as possible. LinkedIn is unique in that it shows you how much of your profile is complete (or not complete) in percentage form. Your goal is to reach the 100% mark.

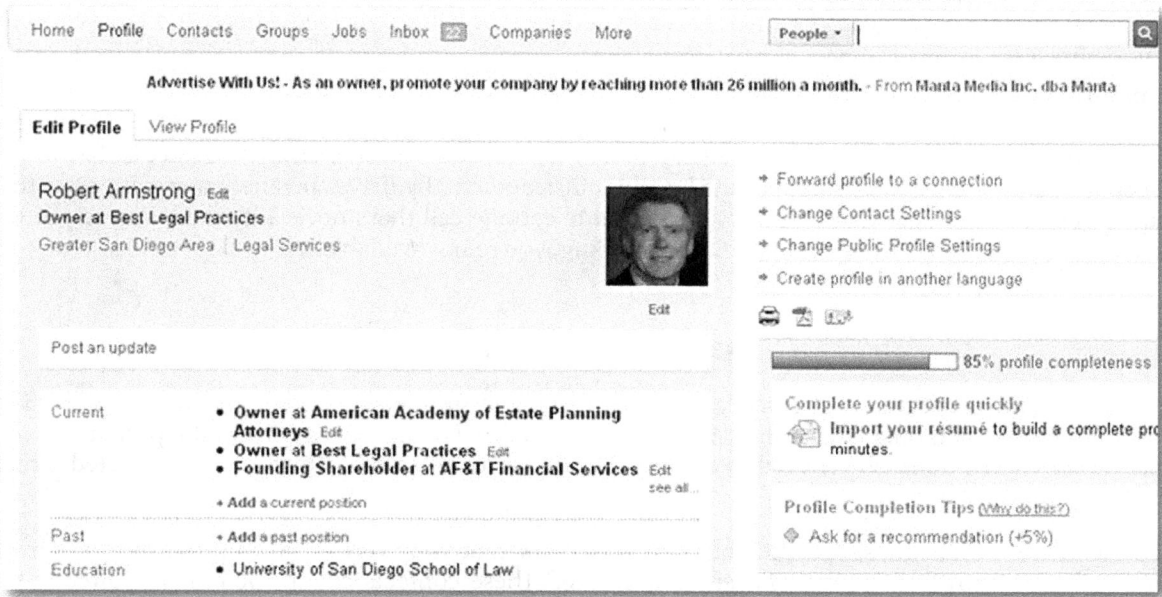

To do this, you need to address the multiple components that make up your profile:

- **Profile Image**—As with other social media platforms, this should be a picture that shows more than just a somber attorney posing for a headshot. If you prefer, upload your firm's logo but if you really want to maximize your profile, use a photo of you instead—just make sure you're smiling.

- **Headline**—The headline is considered to be prime real estate in terms of advertising because it's one of the first things your visitors will see. You have exactly 120 characters to describe yourself, your firm and why people should come to you first. So use this space wisely. In fact, if you're not a writer, consider hiring one to create this blurb for you.

- **Employment**—There are two components here—one for current employment and another for past positions. Obviously, your firm would go in the current position, but be sure to also include some of your past positions to show the extent and range of your experiences.

- **Education**—List the colleges you attend, the degrees you earned and any awards or honors you received in the process.

- **Specialties**—This area is what will determine how you show up in searches within the LinkedIn platform. You want to make sure this section is keyword rich. Think about the services you offer and the clients you'd like to attract. What would they search for?

- **Recommendations**—You must have at least three recommendations before LinkedIn will show your profile as being complete. In order to get recommendations however, you'll need to give a few first

- **Summary**—The summary section is essentially your bio. It's where you can really talk about your services and experience. To keep this section sounding professional, you should write your summary in third person, as if someone else was giving the introduction on your behalf.

- **Groups & Interests**—This section allows you to connect with others who have similar interests and since we've already established that social media is about connecting, it makes sense to utilize this section as much as possible. Each group or interest is a hyperlink that will let you see others who have chosen those same phrases.

A completed profile tells other users two things: 1) Who you are and why they should care and 2) that you're active on LinkedIn. This last one is really important because so many people jump into social media with both feet, only to become disenchanted or distracted and let their accounts go stagnant. The result is that what could have been a quality means of attracting new business instead actually drives business away. It's essentially the equivalent of email messages that go unanswered or a telephone call that's never returned. You've put yourself and your firm out there but when someone comes calling, you're not available.

Building Your Network

The LinkedIn network is actually in three layers. The first is your immediate network or first tier. This includes all of your direct contacts, aka your *"connections"*—people you've done business with in the past and your email contacts. These connections are referred to as your *"first degree"* connections. They can be contacted directly via the LinkedIn email system.

Your connections' connections—people your contacts know that you don't—make up the second tier of your network and are referred to as *"second degree connections."* These contacts cannot be emailed directly because they're not in your immediate network.

Likewise, if you go one tier beyond your second degree network—to the connections of your connections' connections, or third degree connections as it were—you don't have direct access to this group either.

You can, however, contact those in your second and third degree network by using LinkedIn tools called Introductions, InMail or OpenMail.

- Introductions are basically what they sound like. You contact a mutual connection—that is, someone who's in your first degree network as well as the person you'd like to connect with and ask that they make the introduction. Your connection has the option of declining your request but assuming the introduction is made, you can extend this new contact an invitation to join your network. LinkedIn allows five free introductions with a basic (free) account.

- InMail is an email system that allows you to contact any member of LinkedIn without needing an introduction. This service is available only to premium members.

- OpenMail is a messaging system that allows you to receive message from any LinkedIn member without sharing your email address and private information.

Given that you can't directly contact second and third degree network members without an introduction or an upgraded account, how do you go about building and expanding your network of contacts?

There are several ways actually, and we'll cover the most effective ones here.

For starters, you can contact all your colleagues, friends, and other people you know and ask them to join your network. This may not sound beneficial. After all, you already know these people. But hang on! You'll see our logic in a second. The stronger your network is, the more appealing you are to other users so starting with people you already know is a no-brainer. Build your network up as much as possible so that you have something to offer to other users looking to expand their network of contacts.

Now, on to step two: Once you've invited the people you know, join some relevant LinkedIn groups. Yes, LinkedIn has groups and joining some will immediately open you up to new contacts and you can then extend invitations to join your network.

Third, be active. As with other social media sites, the more active you are, the more attention you'll attract. Keep your profile updated. Use LinkedIn apps to pull your blog posts and Twitter tweets into your profile. Add the books you've written, the seminars you're hosting, and anything else you think would be appropriate so that other members can see just how valuable a contact you are. Then sit back and watch the connections pour in!

And lastly, answer questions. LinkedIn has a unique system of allowing users to ask other users questions. This is your chance to be the expert. To get started, choose *"Answers"* from the *"More"* link in the top navigation bar and then click the tab that says *"Answer Questions."*

On the right, you'll see a box that lists all the various categories. There's one for *"Law and Legal"* for example, and, within that category, are several practice areas to drill-down the results even further.

Find categories and questions that you can answer thoroughly and then create your answer. You might even blog about the topic and then provide the link in your posted answer. This is a great way to attract new contacts to your LinkedIn profile while giving your blog a boost at the same time.

MySpace and All the Other SM Sites

While we've covered the top three social media sites, there are, of course, many others. For example, MySpace, once thought to be a general hangout for teens is making a comeback. They've revamped their format in the hopes that they can regain their position in the war of social media sites.

Should you have a MySpace page too?

While this definitely shouldn't be your first priority, it's not a bad idea to have a presence on MySpace. Keep in mind that most of its users tend to migrate to the pages that offer some interactivity—something more than just a company profile. So, if you're going to give MySpace a try, take the time to build a customized page.

In addition to MySpace, there's Foursquare, a location-based social media network that is gaining some popularity. Foursquare encourages its users to *"check-in"* from their mobile devices or via text messaging and will reward points and badges for site participation. This allows users to update their location and connect with friends and followers. It may not sound like a big deal, but given that Foursquare boasts over five million registered users, it certainly doesn't hurt to try and integrate this service into your marketing strategy. For example, you can choose to have your updates automatically tweeted and posted to your fan page as well. Now, imagine that you're hosting an open house or on-site seminar. Wouldn't it be very trendy of you to *"check-in"* with your Foursquare account to update your followers and invite them out to join you?

Flickr is a photo sharing site, also very popular and of course, AVVO is social networking just for attorneys. In addition to these, there are literally hundreds of other sites you may want to consider as part of your social media campaign, keeping in mind that your goal is to connect with potential clients. What better way to do that than to build a presence where those prospects like to congregate?

For a list of various social media sites, see the appendices section of this book.

12.4: A Quick Word on Ethics

With so many easy ways to reach out to your prospects, you might be wondering about the ethical ramifications of social media. And the truth is, the dangers are definitely there.

Social media makes it very easy to set aside the professional standards we've become used to and engage in more casual conversation. But this can get you into serious trouble if you're not careful.

There have been instances where attorneys were reprimanded, sanctioned and even fired for their activities on social media platforms—not because of their mere presence, but because of the actions they took and the statements they made while there.

A North Carolina judge, for example, was reprimanded after *"friending"* an attorney on Facebook. The attorney was appearing before the judge on a case and the two exchanged a few innocent comments about the proceedings online. A public defender in Illinois lost her job for blogging about the cases she was working on, accused of violating the attorney-client privilege and revealing confidential client information.

Does that mean that attorneys can't take advantage of all the tools that online marketing has to offer?

Absolutely not. You can most certainly blog, tweet, and build a rapport with your readers across all the various social media platforms and in fact, it's something you should be doing and doing it regularly.

But attorneys must also be mindful of how these new technologies could potentially clash with the ethics rules that govern our practice. We must take care to ensure that, while our firm learns to be socially-active, it does so without violating the ethics it was founded upon.

Explaining the importance of having a will, filing a trademark, or protecting your assets in divorce is one thing, but if that explanation reveals confidential information or questions the integrity of the judicial system, you might be in for some trouble.

To make matters worse, our ethics rules aren't quite yet ready for all this technology and there are still some *"fuzzy"* areas about what we can and can't do. So much so, that the ABA has formed a work group—the ABA Commission on Ethics 20/20—to study the implications of new technologies on ethics and make formal recommendations to amend the ABA Model Rules of Professional Conduct.

So, what does all of that mean for you?

The strategies and tools outlined in this book are meant to put you in front of your prospects and if you use them correctly, they'll do just that. But regardless of what kind of marketing you do or whether it's online or not, we are still attorneys and we still must answer to the ethics rules that are in place now. Of course, that can vary from state to state. Our best advice is to consult your Bar's ethics rules before launching a social media or online marketing campaign. It could save you major headaches down the road.

Chapter 13:
The Big *Don'ts* of Online Marketing

Okay, now that we've covered most of the things that you should be doing, let's look at some things that you shouldn't even consider.

Often referred to as *"controversial marketing,"* these methods range from the spammy and irritating to the invasive and criminal. But, regardless of their degree of deception, they'll all get you blacklisted by both your readers and your various providers.

Mousetrapping

Mousetrapping is a technique used to keep the visitor on your website, even after he/she is ready to leave. There are several different ways to do this. Some are merely sneaky and frustrating for the user while others are downright illegal. The result is essentially the same: the user tries to *"back-click"* off your site and can't, either because of an array of pop-up windows or because the page has coding in it that causes it to reload infinitely.

This is a big no-no. If your visitor clicks the *"back"* button in the browser, he/she should be able to go back and leave your site.

Spamming

When we hear the word spam, most of us think of the junk email that shows up in our inbox. And this is most certainly one form of spam you should never send. But there is another kind of spam as well. Part of a good online marketing strategy is to get your links posted in social media sites, forums, comments and the like. Unfortunately, some people go overboard and end up alienating their prospective clients instead. One such example involved a prominent law firm in Arizona that decided to blast its marketing message to over 9,000 Usenet discussion groups in 1994. The firm lost their Internet provider as a result because of the large number of complaints received from other Usenet users. This is considered spam, even though your message isn't in email form. So don't do it. There are too many potential ways it could backfire.

Keyword Crazy

We've mentioned this briefly before, but let's address it head-on now. When you're writing your content, write for your reader, not for the search engines. Yes, SEO is important and you should always be mindful of your keyword density but if having the right percentage of keywords makes your content unreadable or hard to understand, you're not going to get the response you're looking for.

Simple Is Best

The average web user reads at about a sixth to seventh grade reading level. So, writing above that is a waste of your time. Instead, keep your content simple and concise. Explain complex topics in laymen's terms and avoid using complicated legal jargon whenever possible.

Link Baiting

Since you're now a full-fledged member of the marketing community, you're going to hear this term quite often in your marketing research. Link baiting is the process of writing a catchy headline to entice users to click through and read your article. In theory, this technique is fine and even recommended as a smart marketing practice. But make sure that your headlines don't promise something your article doesn't deliver. Some marketers use catchy headlines to tempt readers to click in or subscribe, but then don't provide the content implied by the headline itself. This is a sneaky way of doing business and one that should always be avoided.

Part 5:

Bringing It All Together

Chapter 14:
Analyzing Your Results

With so many ways to boost your search engine rankings and connect with your prospects, you might be wondering if there's any way to track how those efforts are paying off.

The answer is yes. Yes, there is.

It has been said that all online marketing can be broken down into two components: traffic and conversion. Another way to explain this is: how many people are seeing your message and how many take action. We're going to measure both.

Now, there are actually a number of different analytical tools but the most popular by far is Google Analytics.

This free tool allows you to track a wide variety of information, including your conversion rates. This is something you'll definitely want to know.

Conversion refers to a user taking a certain action, typically the one you were directing him/her to take. In terms of a law firm's website, this might mean that a user subscribed to your newsletter, enrolled in your webinar or made an appointment to come into your office to retain your firm.

To use Google Analytics, you'll first need to sign up for a free account. Go to **google.com**/analytics to register.

You'll need to log in with your Google ID. So, if you don't have one, sign up for that first. From there, Google will walk you through the set up process. Once you're done, you'll be asked to place a small piece of code on every page in your website you want to track.

This coding won't interfere with your layout and your prospects and clients won't know it's there. Any actions taken on your website, however, will be recorded for you to view and analyze in your Analytics reports.

Google Analytics

Getting Started

Improve your site and increase marketing ROI.

Google wants you to attract more of the traffic you are looking for, and help you turn more visitors into customers.

Use Google Analytics to learn which online marketing initiatives are cost effective and see how visitors actually interact with your site. Make informed site design improvements, drive targeted traffic, and increase your conversions and profits.

Sign up now, it's easy -- and free!

(5M pageview cap per month for non AdWords advertisers.)

Sign Up for Google Analytics

You are just a few steps from Google Analytics. Click on the **Sign Up** button to get started.

Sign Up »

Click the *"Sign Up"* button and then fill in your website's URL, time zone, country and account identifier. Agree to the terms and click *"Submit."* You'll be brought to the Tracking Instructions page that contains the code snippet you need to insert on your website's pages.

Note that this piece of code needs to be on each individual page in order to give you accurate data. To do this, you're going to open up your trusty HTML editor program and insert the entire code just before the </Head> tag:

It will take about twenty-four hours for this code to start generating some data. Once it does, you can monitor your activity from your Analytics dashboard:

From the dashboard, you can see several things, including: the number of visits, the average time your users are staying on your site, your bounce rate, and the percentage that these numbers are changing from week to week. But this is just the beginning.

Click the *"View Report"* link next to your website's URL:

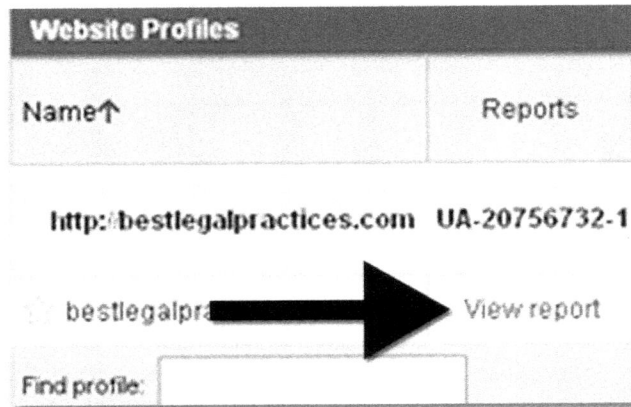

This takes you to the dashboard for this particular site and from here, you can discover a world of information about how your prospects use your site.

You can see activity for each page of content you've created, determine which landing pages are giving you the best results, set up alerts, track your visitors (including mobile visitors, right down to which type of mobile device they're using) and set goals for your site.

This report can be viewed online. You can print it out or you can export it and save it on your hard drive for future reference.

Now, in addition to using Google Analytics, there's one more way you can measure the success of your online marketing efforts: activity. If you're reaching your target market and they like what you're saying, you'll be busier than usual.

There will be more emails to answer, more requests for a consultation, more tweets to acknowledge and more friend requests to approve. You'll see comments on your blog, your tweets will get retweeted and you'll have inquiries from prospects. If you don't experience anything of this, one of two things is happening: either you haven't yet found your target market or you're not getting your message across.

This is why it's so important to measure your results continuously and tweak your campaigns as you go. Try split-testing to see which landing page does better or which offer brings in more appointments. Experiment with different educational reports, different social media sites and different headlines and keywords.

Online marketing is a process of trial and error. So, while we can give you the tools you need and point you in the right direction, your success will depend totally upon your ability to fine tune your campaign and make it your own.

Chapter 15:
Keep the Momentum Going

We've covered quite a bit of ground. So, if you're feeling a little overwhelmed, that's to be expected. After all, you've just taken a huge leap out into the unknown. Chances are you now view your firm very differently from when you first started this book. That can be both exciting and a little unsettling.

The important thing to remember now is to keep your momentum going. You've taken the time to read this book and learn about all the amazing ways you can market your firm. Now, build on that. Don't let your efforts go to waste.

No, you won't be able to create a complex marketing campaign within the next few days. In fact, it's likely that you still won't have everything in place after a few weeks. But that's to be expected too.

There's much to be done and as we've already seen, it's important to do it right. Trying to take everything on at once would be self-sabotage. So go slowly. Start back at the beginning.

Get your website in order. Build your content. Create your offers, your free reports, your newsletters and the like. Build out your social media profiles and get all your third-party services lined up.

Talk with your staff about who will do what because you know you won't be doing all of this alone. Set up your blogging schedule, research groups you'd like to join or directories you'd like to submit to. Figure out what systems you'll need to do what you're wanting to do, and then create them. Trust us when we tell you that systems are everything. Without them, you'll never reach your goals.

Formulate your plan of action first. Decide what it is you want to accomplish and then break that down into smaller goals that aren't so overwhelming. This should be a team effort. Be ready to alter this plan as you go. In the world of online marketing, nothing is written in stone.

Chapter 16:
Do You Need a CMO?

When you founded your firm, you knew you'd be the CEO. You probably also considered the possibility of having a CFO, especially if your firm performed the way you planned.

But, in today's marketplace, there's another position you should be considering as well, especially if you're going to join this digital revolution that's taking place. You need someone who can stay on top of all the emerging technology and keep up with the latest marketing strategies. You need a person on your staff who's devoted to overseeing your social media campaigns, ensuring that goals are being met and that your efforts are paying off.

What you need is a Chief Marketing Officer.

Now, the role of a good CMO is actually more involved than just knowing how to blog and tweet. This person will represent the voice of your ideal client within your firm because it is, in fact, the ideal client that you're targeting. That means that your CMO must be tuned into what's going on in the marketplace. How does your audience respond to your marketing efforts? What would they prefer you do differently?

Remember, you're building a client-focused firm now, one where your client's needs and wants are vital, not just what the attorneys are willing to give. So, your CMO must be able to represent that voice without caving into the desires of the other members of your management team.

Of course, this doesn't mean that your senior partners won't have some say in the final strategies. Quite the contrary! It should be a collaboration of all your various departments. But if the CMO truly represents your ideal client, then that means the rest of your team should be listening intently. If you're not meeting the needs and demands of the CMO, then you're not meeting the needs and demands of your ideal client. That's a losing proposition any way you look at it.

Chapter 17:
What You Need to Know about Outsourcing

By now, you're probably thinking that you're going to need some help if you really want to make this online marketing thing work. And you're right. Mastering all the advancing technology and various strategies takes time, making online marketing a full-time job all by itself.

Now, we've already discussed the importance of having a CMO on staff to oversee your marketing. But, for many small firms, that won't be enough. You'll need to outsource some of the work so that you and your staff can focus on your clients and the practice of law. Before you start contracting third parties, there are a few things you need to know.

The first is that all SEO and SEM companies are not equal. Some companies have limited experience with social media while others aren't really adept at article marketing.

The second thing to know is that very few companies have experience working extensively with law firms and this is a crucial factor in choosing the right company. We've already established that law firms are different from other businesses. So, while you might still need some of the same general services, you're going to need them from a company that knows how to tailor said services to the practice of law.

Track records are another issue you should consider. In addition to having experience in law firm marketing, does the provider you're looking at have a track record of success? It doesn't do any good to publish those articles and post those tweets if you're not reaching your target audience. Your SEO/SEM provider needs to be able to explain how they find your ideal clients and have some sort of measure of success to back up that claim.

Now, the reason we're so familiar with what it takes to market a law firm online successfully is that we've been there. Our SEO/SEM program at the American Academy of Estate Planning Attorneys is successful because we've already done the legwork. We tried all the existing strategies and technologies as we were building our program. So, we know what works and what doesn't. The proof is in the pudding. Academy members dominate their markets online consistently.

We found that, true to form, law firms *are different*. The marketing has to be tweaked *"just so"* and the strategies have to be altered here and there. Your business isn't the average Internet company. So, we had to rethink how to market an attorney's call to action and how to market a firm without sacrificing the prestige that comes with practicing law. Being experienced attorneys ourselves, we also understand the strict Bar ethics rules that control the lives of attorneys.

The writers you hire will need to understand the topics they write about and the people who post your tweets and links and status updates must understand how it reflects upon your image. Your law firm can't be a marketing monster because that's not what a law firm is. Instead, the marketing has to complement and enhance your firm's identity. That's something that not every marketing company will be able to offer.

Just be sure that whoever you hire is qualified to handle your firm the way it should be handled. Don't be afraid to ask questions and if you don't like the answers, keep looking. Bad marketing is actually worse than no marketing at all. So, think carefully before handing your marketing campaign off to someone else.

Now, having said all that, we've got one more exclusive tidbit of information to share. Here it is: We know how much work this will be because we've done it. And we know you're probably wondering how you're going to manage such complex campaigns and still find time to do what you love most: practice law. Well, we've got an answer for that too, so keep reading… There's a special offer in the next chapter just for you.

Exclusive Offer from Best Legal Practices

At Best Legal Practices, we want to make sure that you have everything you need to get the most from your firm's marketing efforts. But we also know that managing a full-scale marketing campaign requires a great deal of time—something that few attorneys have to spare.

So, while this book serves as a great starting point, we realized that what you really needed was a ready-made system… a *"done-for-you"* marketing program that allows you to focus on practicing law while enjoying the benefits that come from strategic online marketing campaigns.

So, that's exactly what we did.

Our SEO/Social Media Marketing Program is divided into two very distinct, very exclusive, packages that include our step-by-step proprietary system for online marketing, a system that we built from the ground up. So we know it works.

These packages offer a wide variety of services, including: blogging, social media posting, backlinking and plenty of tracking and reporting to boot, all designed to help you dominate your market online.

But before we can help you get to where you want to go, you have to know where you are right now. So, the first thing we do is show our clients where they rank in the digital universe. Using several different measurement tools, we give new clients a complete picture of their online presence—their pagerank, their ranking in the search engines, the number of backlinks they have and a variety of other data that allows them to actually see where they stand in the World Wide Web.

And we'd like to offer this analysis to you for free.

As a special thank you for purchasing this book, you can get a complete analysis of your current online presence, presented in hard copy for your convenience as well as a one-hour strategy session with our online marketing experts to go over your analysis and answer any questions you may have. The analysis and your strategy session are completely free. Whether you decide to go forward with our services or not, the analysis report will be yours to keep.

To get your free analysis and complimentary strategy session, visit our exclusive landing page at **bestlegalpractices.com/freestrategy.**

Appendices

We've compiled the following lists of various sites, resources, tools and directories that will help you build your presence on the web. These lists are by no means complete. The web is a constantly changing beast and what's hot today may not even be around tomorrow. Use these lists as a starting point. Then launch your own search to see what's new, what's changed, and what we might have missed.

Legal Directories

*Legal directories are like yellow pages for lawyers. Some of these listings are free while others offer paid solutions only. FindLaw and **Lawyers.com** would be the two most prominent sites on this list. But the others can provide you with increased visibility and assist in getting your site indexed by the search engines.*

Lawyers.com—In addition to a variety of legal forms and related information, **Lawyers.com** includes a full, searchable list of law firm profiles. Start by creating a comprehensive profile for your firm and be sure to include biographies for individual attorneys. To get a free listing on both **Lawyers.com** and their parent company, **Martindale.com** (Martindale-Hubbell), send your basic profile information to listings@martindale.com. If you're interested in an enhanced (paid) listing, you can contact **sales@martindale.com** or go to **www.lawyers.com/legal-marketing/market-your-law-firm.html** and submit the online form. Be sure to also ask about how you can interact with potential clients on the site by participating in the Ask a Lawyer program.

TheLawyersList—Founded in 1904, The Lawyers' List was originally published as a hardbound directory, providing a comprehensive guide of law firms around the world. Today, The Lawyers' List has evolved to include an online search feature as well. You can recommend your firm for inclusion by going to **www.thelawyerlist.net/recommend.shtm**l.

Search-Attorneys.com—Owned by yours truly, this directory consistently ranks at the top of Google and features law firms in the estate planning, elder law, and probate practice areas. Listings start as low as $49 per month and offer guaranteed placement of the first page of results for your city in the directory. To sign up, go to **www.search-attorneys.com/attorneys-Internet-marketing.aspx**.

FindLaw.com—FindLaw is perhaps one of the most prominent legal directories on the web, familiar even to those outside the legal industry. To edit or create a basic profile, you can register at pu.findlaw.com/. FindLaw also offers a variety of enhanced profile features. To learn more, contact them directly at **www.lawyermarketing.com/Company/Contact.shtm**l.

Martindale.com—*See Lawyers.com*

Lawfirm-Directory.com—Lawfirm-Directory allows users to search by location and area of practice. Listings can be obtained for $250 per year and include a backlink to your website.

LawfirmDirectory.org—This site offers four different levels of membership, starting with a budget listing for $299 per year. The budget listing includes basic information about your firm as well as ratings and reviews and links to allow users to add to favorites or email to a friend. At the other end of the spectrum is the Showcase listing for $999 per year. This listing includes images, additional promotional links, coupons, contact forms and a video imbedding widget. Go to **lawfirmdirectory.org/advertise.php** to register and choose your level of membership.

HG Legal Directories—This worldwide directory offers a PR of six and offers two options for adding your firm to the list: a free, basic listing and a premium listing for $195 per year. The basic listing includes contact information about your firm as well as the ability to submit articles to the site's database for public consumption. The premium listing ensures top placement within the directory and also includes a much more enhanced profile, with logos, images and premium listings for all your offices and attorneys as well. To register and choose your listing level, go to **www.hg.org/addlf.html**.

LegalMatch.com—This site is a little different from traditional directories in that it makes an effort to *"match"* you with potential clients. Visitors submit their case information through the site and then matching attorneys are notified and given the chance to review the case details. If you're interested, you can respond to the client and the client can choose to pursue your offer of assistance or not. To get more information, go to **www.legalmatch.com/home/info/infoLP2.do**.

InsuranceLawList.com—This site doesn't actually look like much on the surface but it carries a PR of five and is considered to be one of the top directories used by insurance industry professionals. If this is your practice area, you can list your firm for a one-time fee of $340. Go to **insurancelawlist.com/List_Your_Firm.htm** to complete your listing.

HelpLine Law—HelpLine hosts a worldwide directory of legal services and offers law firms the ability to join on a trial basis for free before opting for the full, paid membership. To obtain your listing, go to **helplinelaw.com/login/lawyer-signup.php**.

Law Research Services—This site offers five different levels of membership, ranging from $8 per month for the basic listing to $50 per month for the top level Sponsor listing. Sponsor listings, include the ability to upload PDFs, PowerPoint presentations and videos as well as top positioning in the search results within the site. To list your firm, go to **www.lawresearchservices.com/MainStreet/list.php**.

Also remember that, in addition to all-inclusive directories, you can also look for industry-specific directories, such as: Insurance Law List mentioned above, as well as local directories within your city, county and state. North Carolina attorneys for example can list on **LawyersNC.com** while firms in San Diego can list locally on **SanDiegoAttorneysOnline.com.** Every listing not only gives you another backlink but also offers the opportunity to increase your visibility online.

Local Directories

(Note: These are in addition to the directories listed in Chapter 9. Effort should first be given to creating your listings in Google Places, Yahoo! Local and Bing Local before focusing on these local directories.)

CitySearch—City Search actually feeds other location-based websites, such as: **Hotels.com** and Expedia as well as MSN. So, while some users don't search on this site directly, their results are fed into some pretty prominent websites. You can list your firm at City Search by going to **www.citysearch.com** and creating a free account. City Search also offers paid advertising so make sure that you know which one you're choosing before you sign up.

Superpages.com—Superpages is a huge online yellow pages directory, complete with social media and mobile presences as well. To list your business for free, go to **superpages.com** and scroll down to the navigation links at the bottom of the page. Click *"Add or Edit a Business"* under the *"About"* section to search for your listing.

Yelp.com—Yelp pulls its data from a variety of sources. So, your firm may already be listed. But that doesn't mean you can't take control of the listing and maximize its effectiveness. Go to **biz.yelp.com** and sign up for a free business owner account to include special offers, reply to review and monitor trends on your listing.

AOL—AOL sponsors a yellow pages directory (YP.com) and you can list your firm for free. Just go to **listings.yellowpages.com/Services/ServiceClaimSearch.aspx** to begin creating your listing. This information will also feed the directory at yellowpages.com.

Local.com—Local.com offers both free and enhanced listings. To set up your free listing, go to **advertise.local.com** and complete the online form. ***Hint: You can preview your listing as you go.***

The enhanced listing gives you the ability to include your logo, special offers, additional photos and other eye-catching extras—depending on the plan you choose, an enhanced listing could start as low as $49 per month. To get your enhanced listing, you'll need to contact **Local.com** directly by calling 888-857-6722 or sending an email with your request to advertisewithus@local.com.

Jayde—Jayde is a popular B2B (business-to-business) search engine with a PR of four. You can get a free listing for your firm by going to **www.jayde.com/submit.html** and completing the online form.

CitySquares—CitySquares offers three levels of membership: free, deluxe and premium. The free profile includes basic info about your firm as well as social media sharing tools, an interactive map, user reviews, coupons and access to basic reporting tools. For $2.49 per month, you can get the Deluxe Profile, which includes all the Basic features as well as the ability to list your hours of operation, include a photo or logo, the removal of ads on your profile page and the ability to privately respond to user reviews. The Premium Profile costs $4.99 per month and comes with all the features we've named so far, plus three additional dofollow links, unlimited photo gallery, video integration, call tracking and the ability to integrate your profile with your account at Constant Contact (if you have one). To upgrade, you'll have to first claim your free listing—go to my.citysquares.com/search to see if your firm is already listed on CitySquares.

Additional local search engines you might want to try include:

- **Yippie.biz—www.yippie.biz/**
- **InfoUSA—www.infousa.com/**
- **Localeze.com— www.localeze.com/**
- **AllPages.com—www.allpages.com/**

- **Kudzu.com—www.kudzu.com/**

- **MagicYellow—www.magicyellow.com/**

- **Openlist—www.openlist.com/**

- **Mojopages—www.mojopages.com/**

- **Hot Frog—www.hotfrog.com/**

- **DMOZ—www.dmoz.org/** (This also feeds the AOL search engine.)

- **SoMuch—www.somuch.com/**

- **LinkCentre—www.linkcentre.com/**

- **Viesearch—www.viesearch.com/**

- **Aviva Directory—www.avivadirectory.com/** (Annual fee of $49.95)

- **Best of the Web—www.botw.org/**

Search Engines

You'll want to focus most of your efforts on the Big Four: Google, Yahoo!, Bing and DMOZ. But submitting your URL to these alternative search engines can help get your site indexed and improve your overall rankings.

- Alta Vista—**www.altavista.com**/cgi-bin/query?pg=addurl

- AllTheWeb—**www.alltheweb.com**/help/webmaster/submit_site

- AOL—data provided through DMOZ (**www.dmoz.org**/)

- BizWeb—**www.bizweb.com**/InfoForm/

- Galaxy—**wwwgalaxy.logika.net**/view/dr_submit.gst?d=

- GigaBlast—**www.gigablast.com**/addurl

- Alexa.com—**www.alexa.com**/siteowners/claim

- ScrubtheWeb.com—**www.scrubtheweb.com**/addurl.html

- SearchHippo.com—**wwwsearchhippo.com**/addlink.php

Blog Directories

Some blog directories operate like basic search engines while others have more of a community feel to them. That means that, for some of these listings, you'll simply submit your blog and be done with it while others will give you the ability to create a profile and interact with other users. As with other types of social media accounts, interaction is key so when you come across these social directories, be sure to include regular updates and postings as part of your online marketing strategy.

BlogCatalog.com—**www.blogcatalog.com**/signup

Bloggapedia.com—**www.bloggapedia.com**/register.php

Bloggernity.com—**www.bloggernity.com**/cgi-bin/add.cgi

Bloghub.com—**www.bloghub.com**/cgi-bin/add.cgi

Bloglines.com—**dashboard.bloglines.com**/signup

Bloggernow.com—**www.bloggernow.com**/cgi-bin/add.cgi

Icerocket.com—**www.icerocket.com**/

Roask.com—**www.roask.com**/suggest-listing.php?id=0

Technorati—**www.technorati.com**

UBDaily.com—**directory.ubdaily.com**/submit.html

URL Shortening Services

*These services "shorten" submitted URLs. This is handy when you have
limited space (such as in Twitter) and when you have an overly-long
link that clutters up your content. All these services are free.*

Bit.ly—Shorten long URLs for free. Customized tags available; analytics provided for registered users. bit.ly

TinyURL—Free; custom tags available, toolbar widget for one-click access. tinyurl.com

Cli.gs—Free registration, analytics, social media monitoring and geotargeting. cli.gs

Ow.ly—Sign in with your Twitter account and you can upload images as well as shorten text URLs. ow.ly

Goo.gl—Google's own URL shortening service, complete with click-through ratios for tracking purposes. goo.gl

DigBig—Free registration, click-through analytics provided. digbig.com

Email Marketing Services

These sites provide web-based email management services. This allows you to create targeted lists for your various marketing campaigns, create autoresponders for your lists, and manage your email marketing campaigns. Some cost more than others. The features can vary as well. So consider how you'll use this service before making your final selection.

iContact—Free 60 day trial, multiple mailing lists, bulk upload. Monthly prices start at $9.99.

ConstantContact—Free 60 day trial, multiple mailing lists, bulk upload. Monthly prices start at $15 per month.

Aweber—First month is $1, multiple mailing lists, bulk upload. Pricing starts at $19 per month.

TrafficWave—30 day free trial. Multiple mailing lists, bulk upload. Pricing is a flat $18 per month fee, regardless of number of subscribers.

MailChimp—Free account up to 1,000 subscribers. After that, pricing starts at $15 per month. Multiple mailing lists and bulk uploads.

Vertical Response—Free 30 day trial. Pricing starts at $10 per month or pay-as-you-go option available. List segmentation and free image hosting.

Social Networking Sites

*(Note: These are in addition to the sites listed in Chapter 12. Effort
should first be given to building your profiles in Facebook, Twitter,
and LinkedIn before focusing on other social media sites.)*

Avvo—Avvo offers both paid and free advertising opportunities, the most popular of which is the free Avvo profile. Like Google Places, your listing already exists—you just have to officially *"claim"* it to start editing. In addition to building a basic profile, you can also showcase your expertise by participating in the Ask A Lawyer program and/or writing your own consumer-oriented legal guides. Upgrade to Avvo Pro for $49.95 per month and you'll also have the ability to add profile taglines, add your Twitter feed and include your blog posts in your profile. To claim your free profile, go to **www.avvo.com**/claim-your-profile?profession_id=1.

Ning—Ning recently did away with their free service but they still offer a unique benefit to your social media campaign—Ning allows you to create your own social media network. With a PR of 7, backlinks from Ning obviously carries some weight and gives you the ability to build your own network of targeted members. You can test-drive Ning free for 30 days by going to **www.ning.com**/chooseplan?o=dm.

Xing—Xing is a social networking platform built specifically for small businesses. Features include a community question application (where you can provide answers if you choose), scheduling and project management tools and Twitter Buzz, a real-time feed of tweets relevant to your practice area and/or location. Xing allows you to share presentations in your profile, manage your projects and connect with other users. Registration is free. Go to **www.xing.com**/app/signup.

Fast Pitch—Fast Pitch is another social networking site devoted to small businesses and professionals. This site is designed to help you promote your professional life, market your firm and build a network of referrals and resources by connecting with other professionals and entrepreneurs. To create your free profile and start connecting, go to **www.fastpitchnetworking.com**/signup.cfm.

EFactor—EFactor is an entrepreneurial community that is focused on connecting small companies and entrepreneurs with potential prospects and investors. Features include discounted member events and travel accommodations, online discussion groups and a virtual marketplace. To create your free profile, go to **www.efactor.com**/signup/.

Social Bookmarking Sites

(Note: These are in addition to the sites listed in Chapter 11. Focus first on building a presence on Digg, Mixx, and StumbleUpon before integrating other bookmarking sites into your strategy.)

- Reddit—www.reddit.com/
- Del.icio.us—del.icio.us/
- Technorati—www.technorati.com/
- Furl—www.furl.net/
- Ma.gnolia—ma.gnolia.com/
- Newsvine—www.newsvine.com/
- Friendfeed—www.friendfeed.com/
- Diigo—www.diigo.com/
- Fark—www.fark.com/
- Dropjack—www.dropjack.com/
- Slashdot—www.slashdot.org/
- Sphinn—www.sphinn.com/

Video Sharing Sites

In addition to YouTube (mentioned in Chapter 11), there are several other popular video sharing websites that will provide you with quality backlinks helping you establish yourself as the expert in your practice area.

- Yahoo Video—video.yahoo.com/
- Google Video—www.video.google.com/
- Metacafe—www.metacafe.com/
- MySpace—vids.myspace.com/
- Photobucket—www.photobucket.com/
- BrightCove—www.brightcove.com/
- DailyMotion—www.dailymotion.com/

- iFilm—www.ifilm.com/
- Flixya—www.flixya.com/
- Viddler—www.viddler.com/
- Lulu—www.lulu.tv/
- GoFish—www.gofish.com/
- Vimeo—www.vimeo.com/
- Pandora—www.pandora.tv/

Podcast Hosting Sites

iTunes—www.itunes.com/

Podbean—www.podbean.com/

Podcast—www.podcast.net/

Podcast Central—www.podcastcentral.com/

Syndic8—www.syndic8.com/

iPod Lounge—www.ipodlounge.com/

PodOMatic—www.podomatic.com/

MyPodcast—www.mypodcast.com/

Zune—social.zune.net/podcasts/

DigitalPodcast—www.digitalpodcast.com/add_anywhere.php?cat=1

Podcast Directory—www.podcastdirectory.com/add/

Podcast Fusion—www.podcastfusion.com/fs/addPodcast.asp

Podcasting Station—www.podcasting-station.com/submitrss.php

Hard Pod Café—www.hardpodcafe.com/modules.php?name=pod_add

Yahoo Media Search—video.search.yahoo.com/mrss/submit?

Podseek—www.podseek.net/

Internet & Online Marketing Lingo

We covered a few of these terms in Chapter 8—terms that you'll run into as you learn more about online marketing. But that's just the tip of the iceberg. As with any industry, online marketing has its own language, so we've decided to include a more comprehensive list here. Just remember that as the industry changes, new terms and definitions evolve, so if you come across a term you don't understand, look it up. A quick search on your favorite search engine will usually do the trick.

- **404**—This refers to an error within your website. When someone clicks on a link pointing to a page that no longer exists, the web returns a 404 error by displaying a generic *"Not Found"* page. You can create custom 404 page—visit our website at **www.bestlegalpractices.com**/law-firm-technology/how-to-create-customized-404-error-pages/ to learn how.

- **Affiliate Marketing**—A program where website owners offer to pay a commission to others to help promote said website.

- **Animated GIF**—This is a type of graphic file that uses a series of static images to create an animated effect.

- **Autoresponder**—A series of emails created in advance and set to send to subscribers on a certain schedule. For example, if a user subscribes to your newsletter, your autoresponders might consist of a welcome email and then several additional emails sent out once a week that educate the reader on additional topics of interest.

- **Below the Fold**—A design term that refers to the space on a given web page below the bottom of the user's screen. In other words, everything that a user CAN'T see without scrolling down.

- **Bookmark**—A link stored in your computer for future reference.

- **Bounce**—This refers to emails sent by you that are returned (bounced) because the email address is invalid.

- **ClickThrough**—The act of clicking on a link contained in your advertising message to the desired destination.

- **Conversion Rate**—The percentage of users who take the desired action, i.e., subscribe to your newsletter, download your report, etc.

- **Cookie**—Information stored on your computer by a website that allows the website to remember your preferences the next time you visit.

- **CPA**—This is an acronym for *"cost-per-action"* and is a type of paid advertising that charges you each time the desired action is completed, i.e., a user completes your submission form or purchases one of your services as a result of the advertisement.

- **CPC**—This acronym stands for *"cost-per-click"* and refers to the amount that the publisher receives each time a user clicks an advertisement displayed on the publisher's website. See PPC for more details.

- **CPS**—CPS is an acronym for *"cost-per-sale"* and refers to the amount that the publisher receives each time a sale is made as a result of the user clicking an advertisement on the publisher's website. This is a more defined version of cost-per-action.

- **CSS**—CSS stands for *"cascading-style-sheets"* and refers to a programming language that defines the look and formatting of a given website. Using style sheets, you can specify and change the look of an entire website by simply changing the code in the style sheet, rather than making changes to each page of the website.

- **CTR**—Your *"click-through rate"* is a way to measure the success of an online marketing campaign. This is calculated by dividing the number of users that clicked on a particular advertisement by the number of times that the ad was actually delivered, also referred to as *"impressions."*

- **Deep Linking**—Linking directly to another webpage other than the site's homepage, i.e., **www.SmithDavisLaw.com/Divorce.html** instead of **www.SmithDavidLaw.com**.

- **Double Opt In**—A method that ensures your subscribers really want to receive your mailings. The user subscribes on your site (known as the *"opt-in"*) and then receives an email asking them to confirm their choice by clicking a link.

- **Ebook**—A book in electronic form, usually a PDF file, but it can also be a self-contained program (executable file) designed to run on your desktop.

- **Ezine**—An electronic magazine or newsletter, delivered via email or through your website.

- **Favicon**—A small icon that's used by browsers to identify a website. This is primarily used as a branding tool as these favicons show up in a user's bookmarks, making your site stand out. To see what a favicon looks like, go to our website at bestlegalpractice.com. Look in the URL window (where you'd type the URL) and you'll see our logo right next to the website's address. Bookmark our site and then you'll see this favicon every time you open your bookmarks folder.

- **Frames**—Before CSS, web designers often used frames to segment their website and thereby control the various elements. A web page could be divided into multiple frames, i.e., leader, left side, right side, main, and each frame represented a separate web page. The home page would then *"call"* the various frames—such as header.html, leftside.html, etc… and compile them together to create the overall look and feel of the site. This allowed web designers to change an element by simply changing the frame page rather than each page of the website. Since CSS however, frames are no longer needed.

- **Guerilla Marketing**—A term coined by Jay Conrad Levinson, this refers to using unconventional marketing methods to get maximum results.

- **Hits**—One of the ways to measure traffic is to look at how many *"hits"* your website received. The term *"hits"* refers to the number of files that have been requested, but not necessarily the number of pages.

- **Homepage**—The main page of a website.

- **HTML**—Hypertext Markup Language. This is the basic coding language that makes webpages look and function the way they do.

- **HTML Banner**—A type of advertising that places a graphic banner on another website. The banner then links to the advertising website.

- **HTML Email**—Email that contains Hypertext Markup Language (i.e., CSS, hyperlinks, etc.), instead of just text.

- **Hyperlink**—A clickable link on a webpage that takes you to another destination.

- **Impressions**—The number of times that a banner or other advertisement is seen by users. This does not mean that the user clicked the ad, only that it was presented to him/her.

- **Inbound Link**—A link pointing to your site from another's website.

- **Joint Ventures**—Just like in the real world, companies can *"partner up"* for a specific promotion. In the online world, a joint venture is where two or more website owners combine forces to increase sales, build their subscriber list or market a particular product or service.

- **Linkbaiting**—Creating headlines that entice users to click through and read the related material. Most often used in social media sites and on blogs.

- **Link Popularity**—Measuring the number and quality of links pointing to your site.

- **Niche Marketing**—Marketing that is targeted to a particular niche or group of people. In relation to your law firm, this would refer to readers only interested in divorce, trademarks, elder law, etc.

- **Organic Traffic**—Traffic that you get for free, meaning that you haven't paid to have your site included in a list or directory.

- **Outbound Link**—A link from your website pointing to another's website.

- **Permission Marketing**—Obtaining customer consent before sending marketing materials.

- **Reciprocal Linking**—Links intentionally traded between two sites.

- **Server**—For purposes of this book, a server is the computer used to host your website.

- **Sig File**—A short paragraph or block of text used at the end of emails to identify the user and provide additional information, such as links to websites, social media profiles, etc.

- **Split Test**—A process of measuring user response by showing one ad, page or email to half of your readers and a different ad, page or email to the other half.

- **Unique Visitor**—Website traffic is measured in hits, views and visits. A visit can occur when a person comes to your site and also when a web robot or server makes a request to your site. A unique visitor (also abbreviated UV) is a human view of your website by someone who has not previously been to your website within the past 24 hours.

- **Viral Marketing**—Marketing messages that entice people to *"pass it along."* This term is used quite often when talking about videos. The subject matter is so controversial or entertaining that those who see it send the link to their friends, who send the link to their friends, and so on, and so on.

- **Webinar**—A seminar that is web-based, i.e., hosted online.

About the Authors

Best Legal Practices is the vision of Robert Armstrong and Sanford M. Fisch, co-founders of the American Academy of Estate Planning Attorneys and co-authors of *The E-Myth Attorney: Why Most Legal Practices Don't Work And What To Do About It.*

Best Legal Practices was founded to provide a comprehensive selection of educational tools and resources to attorneys of all areas of practice. In addition, Best Legal Practices also offers an exclusive SEO/SEM program, helping attorneys achieve online dominance in their preferred markets through strategic Internet marketing practices and social media campaigns.

To learn more about Best Legal Practices and its founders, visit the website at BestLegalPractices.com.

Robert Armstrong

A highly respected attorney, a prosperous legal entrepreneur and a co-founder of two national organizations for estate planning attorneys, Robert Armstrong is committed to making the business-savvy practitioner the standard for professional success in estate planning. His philosophies, systems, and strategies help good lawyers become great, satisfied and successful business owners. He is frequently sought out by attorneys from coast to coast for advice on building successful practices of their own.

The American Academy of Estate Planning Attorneys was inspired by his conviction that attorneys, in order to be successful, need more than just technical legal skills. Today, Robert serves as President of the Academy and is responsible for collaborating on the strategic direction of the organization. On a day-to-day basis, he oversees the Academy's Marketing, Technical Support and Software Development departments. His creative vision, combined with his deep understanding of how to run a successful law practice, generates innovative ideas that make both the Academy, and its members, leaders in the world of estate planning.

Robert is a much sought-after authority on estate planning and is widely quoted in national publications, including: *The Wall Street Journal, Newsweek, U.S. News & World Report, Individual Investor,* and *Money Magazine.*

In addition, he is a co-author of several books, including the practice management book for attorneys, *Creating a Loving Trust Practice,* consumer-centered books, *Total Wealth Management and Estate Planning Basics, A Crash Course in Safeguarding Your Assets.* He also contributed his expertise to the bestseller *Terry Savage on Money.* He is also the co-author of *The E-Myth Attorney: Why Most Legal Practices Don't Work and What to Do about It.*

An articulate spokesman for the advantages of trust-based planning, Robert has been a frequent guest on radio talk shows. He has also appeared on PBS and CBS news programs as an estate planning authority.

Professional Experience
Robert has been a practicing attorney since 1976 and is the founder of one of the most successful estate planning practices in the country. In 1989, he joined forces with fellow estate planning attorney, Sanford Fisch. Together they serve as principals with the San Diego, California law firm, Armstrong, Fisch & Tutoli where they have created thousands of quality estate plans for grateful clients.

In 1993, Robert co-founded the American Academy to help members build successful and robust practices through a rich array of products, services and systems. His innovative, real-world solutions to common professional and business problems help members become the premier estate planning attorney in their communities. With step-by-step coaching, the Academy helps members effectively use technology, successfully and ethically market estate planning services, create accurate accounting systems, inspire law firm staff and stay current with the ever changing estate planning laws.

Education

After four years in the U.S. Navy, including a tour of duty in Vietnam, Robert attended the University of California at San Diego, where he graduated summa cum laude with a Bachelor of Arts Degree in Classical Greek.

In 1976, Robert earned his Juris Doctor degree from the University of San Diego, where he was a staff writer for the Law Review and was awarded the prestigious American Jurisprudence Award for excellence in insurance law.

Robert holds Series 7 and 66 Securities Licenses and a California Life Insurance License so clients receive the benefits of coordinated estate and financial plans.

Sanford M. Fisch

Believing that law is both a rewarding business and a noble profession, Sanford has dedicated himself not only to honing his own legal and technical expertise, but also to helping other attorneys improve their skills and businesses. His ardent desire and deep commitment to helping fellow attorneys drove him to co-found the American Academy of Estate Planning Attorneys.

Today, Sanford continues to serve as Chief Executive Officer at the Academy. He is responsible for collaborating on the strategic direction of the Academy, working with Members on strategic planning in their law firms, increasing firm productivity as well as building alliances with other organizations which provide resources to Academy Members.

By constantly seeking simpler, better and more effective ways of doing things, he continues to make a real difference in the Academy, Member Law Firms, in the lives of families, and the way estate planning law is practiced in the United States.

Sanford is a co-author of several books, including the practice management book for attorneys: *Total Wealth Management and Estate Planning Basics, A Crash Course in Safeguarding Your Assets.* He is also the co-author of *The E-Myth Attorney: Why Most Legal Practices Don't Work and What to Do About It.*

Education

A solid legal education, including a Master's degree in tax law, was the foundation for Sanford's career. He graduated magna cum laude from Boston University in 1977, earning a Bachelor of Science Degree in Economics, where he was a member of the National Mortar Board Honor Society.

In 1980, he went on to earn his law degree from the University of San Diego, where he received the American Jurisprudence Award for excellence in civil procedure. He was also published in the Law Review, *"To Be or Not to Be—Tax is the Question?"*

In 1982, Sanford earned a Master of Laws degree (LL.M.) in Taxation, from Georgetown University Law Center.

Sanford holds Series 7 and 66 Securities Licenses and a California Life Insurance License so clients receive the benefits of coordinated estate and financial plans.

Professional Experience

In addition to building a prominent estate planning law firm, as well as co-founding the American Academy of Estate Planning Attorneys, Sanford has used innovative approaches to help families plan for a comfortable, crisis-free inheritance. Through his work at the Academy, he has become a recognized leader and much sought-after advisor, and consultant to law firms throughout the United States.

From 1982 through 1983, Sanford worked as a tax specialist with Coopers and Lybrand, an international accounting firm. He also taught at the American College for Certified Financial Planners and Chartered Underwriters. He then started and served as principal of the Law Office of Sanford M. Fisch. In 1989, he joined forces with fellow estate planning attorney, Robert Armstrong and served as principal with Armstrong & Fisch, APLC and is currently a principal of Armstrong, Fisch & Tutoli. In 1993, Sanford helped co-found the American Academy of Estate Planning Attorneys.

Notes

Notes

Index

Symbols

.com 41

A

ABA 164
Above the Fold 25
Adwords 41, 152
American Academy of Estate Planning
    Attorneys i, 37, 176
Analytics 169, 171
AVVO 163

B

billing 7, 21
Blog 25, 79, 82, 83, 84, 92, 93, 97, 157,
    185
blogging. *See* Blog
branding 18, 19, 22, 26, 71, 150

C

Chief Marketing Officer. *See* CMO
CMO 175, 176
controversial marketing 165

D

domain name 16, 25, 26, 27, 71, 125

E

email 187
ethics 70, 139, 164, 176

F

Facebook 73, 76, 87, 94, 95, 97, 141,
    142, 143, 144, 145, 150, 151,
    152, 156, 157, 159, 160
Flickr 107, 142, 163
Foursquare 64, 142, 163

G

Google 13, 40, 41, 42, 59, 60, 66, 92,
    94, 95, 99, 142, 152, 159, 169,
    186
Google Analytics 173

Google Places 60

H

Hosting 25, 33
htaccess file 79

I

inbound marketing 12, 15
industry 3, 4, 5, 72
interruption marketing. *See* Inbound
    Marketing
IP address 25

K

keyword density 165
Keywords 15, 37, 38, 77

L

Link baiting 166
LinkedIn 142, 160, 161, 162, 163
long-tail keywords 37, 39

M

Meta Tags 25
microchip 5
mindset 5, 13, 21
mission statement 19, 20, 21
Mousetrapping 165
MySpace 142, 163

N

navigation 25, 72, 77, 80, 97, 145, 150,
    156, 163

O

online marketing i, iv, 8, 11, 13, 15,
    16, 17, 22, 25, 40, 71, 86, 91,
    141, 165, 176
outsource 176

P

Page Rank 37
page views 26

PPC 26, 152
PV. *See* Page Views

S

search-engine marketing. *See* SEM
search engine optimization. *See* SEO
search engines i, 13, 14, 15, 25, 26, 27,
    37, 50, 51, 59, 75, 93, 165
SEM 26, 176
SEO i, 13, 15, 26, 27, 37, 40, 50, 51,
    58, 77, 78, 81, 82, 165, 176
SMM 26
social media i, ii, 8, 15, 16, 26, 76, 83,
    94, 95, 98, 104, 105, 108, 128,
    141, 142, 143, 152, 161, 162,
    163, 164, 165, 175, 176, 186
Social Media Marketing. *See* SMM
social network 160
spam 11, 50, 151, 165
strategic planning 7

T

tag line 18, 19
traditional marketing 6
Twitter 73, 87, 94, 95, 141, 142, 150,
    152, 154, 156, 157, 158, 159,
    160, 163

U

uptime 33
URL 26, 34, 78, 92, 93, 94, 98, 101,
    106, 108, 111, 128, 144, 151,
    155, 156, 160, 186

V

virtual employees 7
vision 18, 19, 20, 21

W

WordPress 34, 76, 77, 78, 79, 92

www.ingramcontent.com/pod-product-compliance
Lightning Source LLC
Chambersburg PA
CBHW061419210326
41598CB00035B/6271